Physics Math Correlations

Small Things and Vast Effects

Thomas J. Buckholtz

T. J. Buckholtz & Associates
Portola Valley, California
USA

Physics Math Correlations
Small Things and Vast Effects

Edition 1

Copyright © 2014 Thomas J. Buckholtz

At the time this edition was published,

- The author could be reached via e-mail at Thomas.J.Buckholtz@gmail.com.
- The author had a website at http://thomasjbuckholtz.wordpress.com.

ISBN: 1502918749
ISBN 13: 978-1502918741
Library of Congress Control Number: 2014919114
CreateSpace Independent Publishing Platform, North Charleston, SC

Printed by CreateSpace Independent Publishing Platform, North Charleston, SC

Table of Contents

Preface

Welcome to *Physics Math Correlations: Small Things and Vast Effects*.

In this book, I propose physics theory. Some of the theory falls beyond traditional physics research. Also, I discuss possibly new mathematics. I hope some of the concepts will prove useful.

For people who think about elementary-particle physics, astrophysics, or cosmology, I hope you can use this book to gain non-traditional interpretations of nature. You can use concepts presented herein to think about fundamental physics.

For people who think about mathematics, I hope you can use this book to gain non-traditional perspective on quantum harmonic oscillators and possibly other areas of mathematics.

For scientists and mathematicians, perhaps you will gain new vantage points for addressing traditional issues. Possibly you will see opportunities to enhance on-going or planned work. Perhaps you will find concepts for other research. Perhaps you will try to verify, refute, or extend work herein.

~ ~ ~

The following contributed to my choosing the title for this book.
- This book discusses attempted physics research. Hence, physics.
- This book discusses attempted math research. Hence, math.
- This book discusses attempts to correlate new or underutilized math solutions with physical phenomena for which traditional math models fall short. Hence, correlations.
- Hence, *Physics Math Correlations*.
- This book features an attempt to develop models correlating with elementary particles and their properties. Hence, small things.
- This book features an attempt to develop models correlating with the rate of expansion of the universe and the shape of the universe. Hence, vast effects.
- Hence, *Physics Math Correlations: Small Things and Vast Effects*.

~ ~ ~

I hope people will use this work. I hope people will benefit from this work. I hope people will tell me of extensions to this work, shortcomings in the work, and developments to which the work contributes.

- Thomas J. Buckholtz

Portola Valley, California USA
November 2014

Dedication

To Helen Buckholtz
And, in memory of Joel and Sylvia J. Buckholtz

About the author

Dr. Thomas J. Buckholtz is the author or a coauthor for articles, books, chapters, or reports regarding physics, applied physics, mathematics, computer science, applied computing, computer-based games, software licensing, innovation, systems-thinking tools, the information age, information proficiency, service science, governmental service to the public, and roles of chief information officers.

He played pivotal roles in the following endeavors.

- Create lines of business for a $1 billion (annual revenue) business unit.
- Save $100 million per year for a $6 billion company.
- Pioneer three information technologies.
- Establish three information-technology marketplace business practices.
- Develop useful business, engineering, and scientific software.
- Double a two-person firm's revenue, for each of two consecutive years.
- Preserve 7 kilometers of Pacific Ocean coastline.
- Create an international service program.
- Improve governmental service (from all levels of government) for the American public.
- Create a grassroots line-of-business for a United States political party's National Committee.

Tom served in the following capacities.

- Executive leading a $1 billion business unit
- Corporate officer and advisor for startups
- Chief information officer (CIO) for a $10 billion enterprise
- Co-CIO for the United States federal government's Executive Branch
- Program leader advocating innovation, enhancing teamwork, and providing information technology throughout a $6 billion company
- Commissioner, United States General Services Administration
- Mathematician; Physicist; Scientist; Engineer
- Professorial Lecturer; University Extension Instructor
- Speaker; Workshop provider; Author
- Business advisor; Innovation consultant

Dr. Buckholtz's clients and employers have included large and small enterprises in aerospace, agricultural research, biotech, business services, computing, defense, education, energy utilities, government, healthcare, high technology, innovation, insurance, Internet, law enforcement, politics, research and development, telecommunications, and venture capital.

Tom served on elected or appointed boards or in other volunteer capacities for a residential cooperative, a swim club, and organizations in academia, innovation, and public policy. For a successful United States presidential campaign, he served as a donor, fundraiser, policy-research committee member, speaker, alternate delegate at the candidate's party's National Convention, speakers bureau leader, county cochairman, and county representative at regional and statewide meetings. He served as co-producer and co-host for 250 interview-format television programs discussing business, charitable, community, educational, governmental, and political endeavors.

His education includes the following.

- Earn a B.S. in mathematics from the California Institute of Technology.
- Earn a Ph.D. in physics from the University of California, Berkeley.
- Complete business administration programs at Stanford University and the University of Michigan.

Notice

Small Things and Vast Effects

Thomas J. Buckholtz

Thomas.J.Buckholtz@gmail.com

Part 0   Introduction and summary

Section 0.1   Abstract

What elementary particles remain not found? What phenomena would their existence explain? We try to answer those questions.

We use the following steps. Find math for which some solutions correlate with known elementary particles and with interactions in which the particles partake. Assume that other solutions correlate with undiscovered particles and interactions. Consider the new particles and interactions. Address known particle-physics problems. Address known astrophysics problems. Address known cosmology problems.

Solutions point to dark-matter and dark-energy fermions. Solutions point to other particles. New particles may cause symmetry violations. New bosons may affect the rate of expansion of the universe. New particles may provide for other aspects of particle-physics, astrophysics, and cosmology.

Solutions correlate with particle properties. For neutrinos, we predict masses and Dirac-or-Majorana fermion types. For leptoquark-like particles, we predict charges, masses, and minimum numbers of particles in particle clusters.

The math features isotropic quantum harmonic oscillators. The math provides the solutions. The math provides a basis for quantum theory. The quantum theory correlates with phenomena for which people associate models based on general relativity.

Section 0.2   Summary

In Part 1, we catalog all known elementary particles, the graviton that people hypothesize, and some other possible particles. In this part, we do not discuss dark-matter elementary particles and we do not discuss dark energy.

In Part 2, we suggest ways to resolve problems that people associate with the concept that traditional theoretical physics does not adequately correlate with observed phenomena. People might associate the problems with the areas of elementary-particle physics, astrophysics, and cosmology. Topics include the rate of expansion of the universe, curvature (if any) of the universe, baryon asymmetry, violations of CPT-related symmetries, quasars, and the number (3) of generations of fermions. Also, we predict a tauon mass $(1.776814(\sim48)\times10^3 \text{ MeV}/c^2)$ with a smaller uncertainty range than experiments have yielded.

In Part 3, we provide mathematics that underlies much of this paper.

In Part 4, we predict masses {or mass-math eigenvalues} for neutrinos, of which 2 would be Majorana fermions (~0.125 eV/c^2 and ~0.058 eV/c^2) and 1 would be a Dirac fermion (~0.058 eV/c^2). We predict masses (from 80.4 GeV/c^2 to 113.7 GeV/c^2) and lower bounds for cluster-production threshold energies (from ≥241 GeV to ≥569 GeV) for O-family elementary particles. We discuss the applicability of concepts of invariant and relative.

In Part 5, we describe elementary particles that could provide bases for dark matter and for dark-energy stuff. We also describe possible siblings (related to known leptons) that people might call elementary particles.

In Part 6, we offer seemingly straightforward ways to discuss spin and to discuss helicity, chirality, and/or handedness. Also, we correlate with our quantum-based approach various aspects of conservation of energy, momentum, and angular momentum; photonics; and general relativity.

In Part 7, we provide quantum-based models correlating with planetary perihelia shifts and other phenomena people traditionally model via general relativity. Also, we discuss inflation, asymptotic freedom, the lack of observing free quarks or gluons, black-hole thermal radiation, a force that helps hold galaxies together, the spacecraft flyby anomaly, the extent to which neutrinos are Dirac fermions or Majorana fermions, vacuum zero-point energy and the cosmological constant, and potential correlations with nuclear physics.

In Part 8, we discuss some observational and theoretical topics not addressed above. An example of an observational topic is observed ratios of densities for dark energy and non-dark-energy. An example of a theoretical topic is extra dimensions. We point to possibilities for a perturbation theory.

In Part 9, we list changes (that people might say our work suggests) to traditional narratives. The narratives involve areas of mathematics, mathematical physics, elementary-particle physics, astrophysics, and the evolution of the universe (cosmology). For each area, we also note possible opportunities for research.

Section 0.3   Keywords

Astrophysics, Asymptotic freedom, Baryonic matter, Black-hole thermal radiation, Boson, Charge, Chirality, Clumping, Color charge, Conservation laws, Cosmic microwave background (CMB), Cosmology, CPT symmetry, Curvature of the universe, Dark energy, Dark matter, Density of the universe, Dirac neutrino, Elementary particles, Extra dimensions, Fermion, Field theory, Flat universe, Flavor, Flyby anomaly, Free quarks, Fundamental forces, Galaxy rotation problem, General relativity, Generations, Gluon, Gravitational constant, Graviton, Handedness, Helicity, Higgs boson, Inflation, Lasing, Leptoquark, Magnetic moment, Majorana neutrino, Mass, Masses of elementary particles, Mathematical physics, Matter/antimatter imbalance, Neutrino masses, Photon, Quantum gravity, Quantum harmonic oscillator, Quasar, P violation, Perihelion shift, Rate of expansion of the universe, Space-like behavior, Special relativity, Spin, Standard Model, Standard Model symmetry, Strong interaction, Tauon mass, Time-like behavior, Uncertainty, Unified electromagnetism and gravity, Vacuum zero-point energy, Vector potential, Weak interaction

Section 0.4   Guesses

This work includes guesses.

The next statement provides a description of how science works made by Richard Feynman. [Ref.0.4.1]

> In general we look for a new law by the following process. First we guess it. Then we compute the consequences of the guess to see what would be implied if this law that we guessed is right. Then we compare the result of the computation to nature, with experiment or experience; compare it directly with observation, to see if it works. If it disagrees with experiment it is wrong.

An approach to the research in this paper features activities that include the following. Guess at a concept for an area to explore. Develop tentative models or theory. Review models or theory with regard to observations or experiments, traditional vocabulary and statements, and aspects of this work. Write tentative results into a draft paper covering aspects of the research. Review the research for consistency. Review the draft for readability and consistency. Go back to fix problems. Go forward regarding other areas.

~ ~ ~

Ref.0.4.1   John R. Gribbin and Mary Gribbin, *Richard Feynman, A Life In Science*, Dutton, 1997, page 178.

Section 0.5   Structure

The next table notes places where this paper performs various functions. The place column indicates the section that performs the activity the function column notes. The list column shows letters that partly label some statements in this paper. This paper includes statements with labels of the forms X j.k.n and X.j.k.n. There, X is 3 or more letters, j is a part number, j.k is a section number, and n is a number. This paper lists references regarding numeric data.

Table 0.5.1   Sections that collect lists

Function	Place	List	
			(0.5.1)
• Summarize results	Section 10.1	Abs	(0.5.2)
• And, estimate some context for results			
• List guesses	Section 10.2	Gss	(0.5.3)
• List possible opportunities for research	Section 10.3	Tbd	(0.5.4)
• List names of tables	Section 10.4	Table	(0.5.5)
• List references	Section 10.6	Ref	(0.5.6)

In conjunction with Abs statements, we provide statements that follow the symbol ▪. People might read an item that begins with a ▪ as "people might say that ... {text that follows the ▪}." We think that each of some of the statements preceded by ▪ point to possible opportunities for research.

The next items pertain.

Tbd.0.5.1   For each statement following a ▪, to what extent does the statement   (0.5.7)
point to possible opportunities for research?

Tbd.0.5.2   For each statement labelled Gss, to what extent does the statement   (0.5.8)
point to possible opportunities for research?

Part 1   Elementary particles (not including dark matter or dark energy)

Section 1.0   Introduction and summary

Abs.1.0.1   We list 7 families of elementary particles.
   ▪   We report results from using a math model to catalog elementary particles.

In Part 1, we catalog all known elementary particles, the graviton that people hypothesize, and some other possible particles. In this part, we do not discuss dark-matter elementary particles and we do not discuss dark energy.

The next table provides bases for cataloging elementary particles. [Part 3; for example, Section 3.4]

Table 1.0.1   Parameters that correlate with properties of elementary particles
   •   For an elementary particle, exactly 1 of the following pertains for S, which   (1.0.1)
        denotes spin/ℏ
      •   2S is a non-negative even integer
          •   The term boson pertains for such a particle
      •   2S is a positive odd integer
          •   The term fermion pertains for such a particle
   •   For an elementary particle, exactly 1 of the following pertains for a   (1.0.2)
        parameter Ω
      •   S>0 and $\Omega = +S(S+1)$
      •   S=0 and $\Omega = 0$
      •   S>0 and $\Omega = -S(S+1)$
   •   For an elementary particle, m denotes mass and exactly 1 of the following   (1.0.3)
        pertains
      •   m = 0
      •   m ≠ 0

Items above correlate with possibilities for 10 families of elementary particles. Here, 10 = 2 + 8. The first term (2=1×1×2) corresponds to S=0. The second term (8=2×2×2) corresponds to S≠0. Observations have yet to detect (and models this paper presents do not seem to correlate with) the 1 possibility for which S=0 and m=0. Observations have yet to detect (and models we present do not seem to correlate with) the 2 possibilities for which 2S is odd and m=0. We discuss 7 (=10−1−2) families.

The next table defines symbols for 7 families. Each known elementary particle correlates with a family. The examples column lists some known particles and the (hypothetical) graviton.

Table 1.0.2   7 families of elementary particles

2S	Ω	Mass	Traditional theme	Symbol for the family	Examples of elementary particles	(1.0.4)
Even	>0	0		G	Photon, graviton	(1.0.5)
Even	>0	≠0	Weak	W	Z, W⁻, W⁺	(1.0.6)
Even	0	≠0	Higgs	H	Higgs	(1.0.7)
Even	<0	≠0		O		(1.0.8)

2S	Ω	Mass	Traditional theme	Symbol for the family	Examples of elementary particles	(1.0.4)
Even	<0	0	Strong	Y	Gluon	(1.0.9)
Odd	>0	≠0	Leptons	L	Electron, muon	(1.0.10)
Odd	<0	≠0	Quarks	Q	Up, anti-up	(1.0.11)

In this part, we predict elementary particles correlating with the G- and O-families. In subsequent parts, we discuss possible elementary particles correlating with the O-, Y-, L-, and Q-families.

Section 1.1   Photons

Abs.1.1.1   We use 4 harmonic oscillators to describe some aspects of photons.
  ▪   This work provides a way to discuss observer-invariant aspects of photons.

Suppose a source emits circularly polarized photons. Suppose that just 1 ground-state energy characterizes all the emissions. Suppose that multiple observers can detect these emissions. (Each emission can be observed by only 1 observer.)

The next table shows properties upon which all observers would agree. The symbol a^+ denotes a raising operator. The symbol a^- denotes a lowering operator. Item (1.1.5) defines Œ for photons.

Table 1.1.1   Invariant properties related to photons
  • N(P2L) denotes the number of times the left circular polarization mode is   (1.1.1)
    excited
    • N(P2L) is a non-negative integer
  • N(P2R) denotes the number of times the right circular polarization mode is   (1.1.2)
    excited
    • N(P2R) is a non-negative integer
  • N(E0) = N(P2L) + N(P2R)   (1.1.3)
    • N(E0) is a non-negative integer
    • People correlate N(E0) + 1/2 with energy
  • N(P0) denotes a quantum number correlating with longitudinal   (1.1.4)
    polarization
    • N(P0) = −1
    • a^+ | N(P0)=−1 >
      • = $(1+\{N(P0)=-1\})^{1/2}$ | N(P0)=0 >
      • = $(1+(-1))^{1/2}$ | N(P0)=0 >
      • = 0 | N(P0)=0 >
    • N(P0) = −1 correlates with no longitudinal excitation occurring
  • Œ = (N(E0) + 1/2) − { (N(P0) + 1/2) + (N(P2L) + 1/2) + (N(P2R) + 1/2) }   (1.1.5)
  • Œ = 0   (1.1.6)

- For physics-relevant mathematics solutions, at least 1 of N(P2L) and   (1.1.7)
  N(P2R) is 0
  - If each of N(P2L) and N(P2R) were positive, people could subtract the
    minimum of N(P2L) and N(P2R) from each of the 2 numbers to obtain a
    physics-relevant statement
  - In other words, a unit of left circular polarization and a unit of right
    circular polarization sum to 0
- For a discussion limited to photons, $a^- \mid N(P0)=-1 >$ is irrelevant   (1.1.8)
  - The combination of item (1.1.7), Œ=0, N(E0)≥0, N(P2L)≥0, and N(P2R)≥0
    prohibits this lowering-operator operation
- People can associate each of the 4 N(j) with a harmonic oscillator, j   (1.1.9)
- People can consider that the set of 4 harmonic oscillators includes 2   (1.1.10)
  isotropic (that is, equal strength) sets, each consisting of 1 or more
  isotropic harmonic oscillators
  - The QE-like set consists of oscillator E0
    - People might associate the term QE-like with the terms energy-like or
      time-like
  - The QP-like set consists of oscillators P0, P2L, and P2R
    - People might associate the term QP-like with the terms momentum-
      like or space-like
- The raising operator for each of the 2 P2j oscillators satisfies the following   (1.1.11)
  - $a^+ \mid N(P2j) > \ = \ (1+ N(P2j))^{1/2} \mid N(P2j)+1 >$
- S = 1   (1.1.12)
  - S denotes spin/ℏ
- c = E / P   (1.1.13)
  - E denotes the observed energy
  - P denotes the observed momentum

Item (1.1.4) shows a negative quantum number for a harmonic oscillator. We assume the algebraic factor $(1+N)^{1/2}$ (in which N is a harmonic oscillator quantum number) pertains for N<0. People might think that the factor does not apply because states for N<0 cannot exist. We think that, for a QP-like isotropic multidimensional harmonic oscillator with D dimensions, the least value of N that applies for models that correlate with elementary bosons is the smallest integer greater than or equal to −D/2. [Part 3]

The next table notes properties about which various observers may disagree. Here, various observers could assign different values to the ground state energy ξ'. Here, people might say that E=ξ and P=ξ/c.

Table 1.1.2   Relative properties related to photons
- ξ and ξ' in the expression $\xi = \xi' (N(E0) + 1/2)$ (1.1.14)
- The perceived direction of motion   (1.1.15)

The next table shows ground states for 2 photon modes. The first mode has left circular polarization. The second mode has right circular polarization. To form the j in the N(j) correlating with a column, append the second row in the column to the first row in the column. For example, the leftmost relevant oscillator is oscillator E0. The rightmost relevant oscillator is P2R. A # denotes a 0 that does not change for the mode.

Table 1.1.3   Photon modes and their ground states

E6R	E6L	E4R	E4L	E2R	E2L	E0	P0	P2L	P2R	P4L	P4R	P6L	P6R	P8L	P8R	Particle or Mode	
						0	–1	0	#							22G2L	(1.1.18)
						0	–1	#	0							22G2R	(1.1.19)

(1.1.16)
(1.1.17)

The next table shows the Nth excited state for each mode.

Table 1.1.4   Photon modes and their Nth excited states

E6R	E6L	E4R	E4L	E2R	E2L	E0	P0	P2L	P2R	P4L	P4R	P6L	P6R	P8L	P8R	Particle or Mode	
						N	–1	N	#							22G2L	(1.1.22)
						N	–1	#	N							22G2R	(1.1.23)

(1.1.20)
(1.1.21)

~ ~ ~

Work above regarding photons features 4 harmonic oscillators. People can associate 4 energy-momentum space coordinates - p0, p1, p2, and p3 - with the photon representation. People might associate 4 space-time coordinates - x0, x1, x2, and x3 - with the 4 energy-momentum space coordinates. Here, the space-time coordinate-x1 axis parallels the motion of the photon. The next table pertains.

Table 1.1.5   Coordinates relevant for photons
- p0 associates with oscillator E0   (1.1.24)
- p1 associates with oscillator P0   (1.1.25)
- p2 associates with oscillators P2L and P2R via $(22G2L + 22G2R)/2^{1/2}$ (1.1.26)
- p3 associates with oscillators P2L and P2R via $(22G2L - 22G2R)/(\{-i\}2^{1/2})$ (1.1.27)
- x0 associates with p0   (1.1.28)
- x1 associates with p1   (1.1.29)
- x2 associates with p2   (1.1.30)
- x3 associates with p3   (1.1.31)

The next table correlates coordinates with the terms QE-like and QP-like.

Table 1.1.6   QE-like coordinates and QP-like coordinates

Term	Energy-momentum space coordinates	Space-time coordinates	
QE-like	p0	x0	(1.1.33)
QP-like	p1, p2, p3	x1, x2, x3	(1.1.34)

(1.1.32)

Section 1.2   Gravitons

Abs.1.2.1   We use 8 harmonic oscillators to describe aspects of gravitons.
- This work provides a quantum description for gravity.

People find difficulty in expressing effects of gravitation via a flat space-time coordinate system that includes no more than 1 QE-like coordinate and no more than 3 QP-like coordinates. People say that, if gravitons exist, gravitons have 2 polarizations. We anticipate adding 4 oscillators - E2R, E2L, P4L, and P4R - in order to describe gravitons. The next items pertain.

Gss.1.2.1   Oscillators P2L and P2R correlate with charge and with spin-1.   (1.2.1)

Gss.1.2.2   Oscillators P4L and P4R correlate with mass, with spin-2, and with 2   (1.2.2)
      polarizations for gravitons.

Gss.1.2.3   For gravitons, N(P2L) = N(P2R) = #.   (1.2.3)

Gss.1.2.4   For the purposes of this work, the 2 P4j oscillators need not   (1.2.4)
      correlate (ultimately) with additional QP-like space-time coordinates.

- We are considering circular polarization
- We can choose coordinates such that …
  - p2 associates with oscillators P4L and P4R
    - via $(44G4L + 44G4R)/2^{1/2}$
      - See Table 1.1.5 and Table 1.2.1
  - p3 associates with oscillators P4L and P4R
    - via $(44G4L - 44G4R)/(\{-i\}2^{1/2})$
      - See Table 1.1.5 and Table 1.2.1

Gss.1.2.5   Oscillators E2R and E2L pertain.   (1.2.5)

- This correlates with people's perceived needs for more than 4 flat space-time coordinates and with item (1.2.4)

Gss.1.2.6   N(Ej) = 0 (for j = 2R, 2L, and 0) correlates with graviton ground   (1.2.6)
      states.

- Œ = { (N(E2R) + 1/2) + (N(E2L) + 1/2) + (N(E0) + 1/2) } − { (N(P0) + 1/2)   (1.2.7)
+ (N(P2L) + 1/2) + (N(P2R) + 1/2) + (N(P4L) + 1/2) + (N(P4R) + 1/2) }

Gss.1.2.7   Œ = 0 for gravitons.   (1.2.8)

- N(P0) = −1 for gravitons   (1.2.9)
  - This follows from Œ=0

$\sim \sim \sim$

We explain the notation jkG… . Here, j denotes #P. Here, the symbol #P correlates with the largest n in an applicable N(Pnl). Here, l denotes L, R, or blank. More generally, #P+1 denotes the number of relevant QP-like oscillators. Here, k denotes 2S. G denotes the G-family. We select the name G-family to correlate with G being the first letter in gamma ray (a type of light) and the first letter in gravity.

Also, we define a symbol #E such that #E+1 denotes the number of relevant QE-like oscillators.

The next table shows ground states for graviton modes. We use the symbol @ to denote 0 for cases in which a representation of particles points to the possible existence of multiple instances of particles. Regarding gravitons, we show (in Part 3 and Part 5) that the universe may include 8 instances of gravitons. The number 8 correlates with the number of generators for an SU(3) symmetry. The SU(3) symmetry correlates with the 3 harmonic oscillators E2R, E2L, and E0.

Table 1.2.1   Graviton modes and their ground states

E 6R	E 6L	E 4R	E 4L	E 2R	E 2L	E 0	P 0	P 2L	P 2R	P 4L	P 4R	P 6L	P 6R	P 8L	P 8R	Particle or Mode	
				@	@	@	−1	#	#	0	#					44G4L	(1.2.12)
				@	@	@	−1	#	#	#	0					44G4R	(1.2.13)

(1.2.10)
(1.2.11)

Section 1.3   The G-family

Abs.1.3.1   The G-family includes all zero-mass elementary particles except Y-family particles.
- This work unifies quantum electromagnetism and quantum gravity.

The G-family includes only zero-mass bosons. (The G-family does not include gluons and other possible Y-family elementary particles.)

Above, we discuss the G-family members 22G2& (photon) and 44G4& (graviton). In this notation, the symbol & replaces the collection of L symbols and R symbols. Notation of the form jkG%& (for j=#P, for k=2S, and for some list % of small non-negative even integers) denotes a particle. Each G-family particle has 2 modes.

The next table shows ground states for possible G-family modes. We denote the corresponding particle by 42G24&. The modes have spins of 1, based on |S(P4n")−S(P2n')| = |2−1|, with S denoting spin and with n" not equal to n'. (For example, for 42G2R4L, n"=L, S(P4L)=2, n'=R, and S(P2R)=1.) Paralleling discussion above, these modes correlate with 1 QE-like space-time coordinate and 3 QP-like space-time coordinates.

Table 1.3.1   42G24& modes and their ground states

| E | E | E | E | E | E | P | P | P | P | P | P | P | P | P | P | Particle or | |
6R	6L	4R	4L	2R	2L	0	0	2L	2R	4L	4R	6L	6R	8L	8R	Mode	
						0	−2	#	0	0	#					42G2R4L	(1.3.3)
						0	−2	0	#	#	0					42G2L4R	(1.3.4)

(header rows labeled (1.3.1) and (1.3.2))

The next table shows first excited states for 42G24& modes. Here, regarding item (1.3.9), N(mode)=1.

Table 1.3.2   42G24& modes and their first excited states (N(mode) = 1)

| E | E | E | E | E | E | P | P | P | P | P | P | P | P | P | P | Particle or | |
6R	6L	4R	4L	2R	2L	0	0	2L	2R	4L	4R	6L	6R	8L	8R	Mode	
						2	−2	#	1	1	#					42G2R4L	(1.3.7)
						2	−2	1	#	#	1					42G2L4R	(1.3.8)

(header rows labeled (1.3.5) and (1.3.6))

Throughout the G-family, the next item describes raising operators.

- $a^+ | N(mode) > = (1+ N(mode))^{n/2} | N(mode)+1 >$ (1.3.9)
  - $n = -N(P0)$ = the number of QP-like oscillators for which 0 (and not #) pertains for the ground state

The next item pertains.

- For G-family particles and modes for which N(P0)=−1, 2S = #P   (1.3.10)

The next table shows modes for some S=2 G-family elementary particles (other than gravitons). We denote the 2 particles (that the table shows) by 64G246& and 84G2468&, respectively. For example, for the 64G246& particle, #E=0 and #P=6.

Table 1.3.3   S=2 #E=0 G-family modes, and their ground states

| E | E | E | E | E | E | P | P | P | P | P | P | P | P | P | Particle or | |
6R	6L	4R	4L	2R	2L	0	0	2L	2R	4L	4R	6L	6R	8L	8R	Mode	
						0	−3	0	#	#	0	0	#			64G2L4R6L	(1.3.13)
						0	−3	#	0	0	#	#	0			64G2R4L6R	(1.3.14)

(header rows labeled (1.3.11) and (1.3.12))

E	E	E	E	E	E	E	P	P	P	P	P	P	P	P	P	Particle or	(1.3.11)
6R	6L	4R	4L	2R	2L	0	0	2L	2R	4L	4R	6L	6R	8L	8R	Mode	(1.3.12)
						0	−4	#	0	0	#	#	0	0	#	84G2R4L6R8L	(1.3.15)
						0	−4	0	#	#	0	0	#	#	0	84G2L4R6L8R	(1.3.16)

We call the series 22G2&, 42G24&, 64G246&, and 84G2468& the G-family #E=0 series. From work above, this series correlates with 1 QE-like space-time coordinate and 3 QP-like space-time coordinates.

The next table shows G-family #E=2 particles. For each particle jkG%&, exactly 2 modes for which k=2S exist. Item (1.3.19) shows the graviton.

Table 1.3.4 #E=2 G-family particles and their ground states

E	E	E	E	E	E	E	P	P	P	P	P	P	P	P	P	Particle or	(1.3.17)
6R	6L	4R	4L	2R	2L	0	0	2L	2R	4L	4R	6L	6R	8L	8R	Mode	(1.3.18)
				@	@	@	−1	#	#	0	0					44G4&	(1.3.19)
				@	@	@	−1	0	0	#	#					42G2&	(1.3.20)
				@	@	@	−2	#	#	0	0	0	0			62G46&	(1.3.21)
				@	@	@	−2	0	0	#	#	0	0			64G26&	(1.3.22)
				@	@	@	−2	0	0	0	0	#	#			62G24&	(1.3.23)
				@	@	@	−3	#	#	0	0	0	0	0	0	82G468&	(1.3.24)
				@	@	@	−3	0	0	#	#	0	0	0	0	84G268&	(1.3.25)
				@	@	@	−3	0	0	0	0	#	#	0	0	82G248&	(1.3.26)
				@	@	@	−3	0	0	0	0	0	0	#	#	84G246&	(1.3.27)

In Table 3.4.14, we note some mathematical solutions that possibly correlate with G-family members with #E>2.

The next table lists G-family particles for which we find the most significant roles below. For many purposes, the effect of each particle (other than 44G4&) in Table 1.3.4 would be much smaller than the effect of some particle in Table 1.3.5.

Table 1.3.5 Selected G-family members

- 22G2&, 42G24&, 64G246&, 84G2468& (1.3.28)
  - These are the members of the G-family #E=0 series
- 44G4& (1.3.29)
  - This is a member of the G-family #E=2 set
  - This the graviton

Section 1.4 Results from math correlating with non-zero-mass elementary particles and their fields

Abs.1.4.1 We summarize math solutions that we correlate with non-zero-mass elementary particles.

- This work provides insight into concepts of particles and fields.

Part 3 provides mathematics leading to solutions that may correlate with elementary particles. The math features isotropic quantum harmonic oscillators. Here, we summarize some findings.

The math features ν, a parameter that, for non-zero-mass elementary particles, correlates with 2 concepts. One of the concepts is a choice between fermion and boson. The other concept is a choice among elementary particle and/or field related to elementary particles.

Table 1.4.1   A parameter (ν) that correlates with particles and fields for non-zero-mass particles
- $\nu = -1/2$ correlates with fermion fields   (1.4.1)
- $\nu = -1$ correlates with boson fields and particles   (1.4.2)
- $\nu = -3/2$ correlates with fermion particles   (1.4.3)

The math also features the 3 parameters Table 1.0.1 lists.

Regarding solutions that correlate with elementary particles and/or their fields, the math also features a parameter D. The next table pertains. For positive values for Ω, the limit D>0 places upper limits on the size of Ω for non-zero-mass elementary particles. The limit D>0 does not place an upper limit on the size of $|\Omega|$ for $\Omega<0$. Section 2.7 discusses the number of generations. The term Majorana fermion denotes a fermion that is its own antiparticle. In contrast, the term Dirac fermion denotes a fermion that has a distinct antiparticle. For fermions, item (1.4.7) shows an upper limit on the number of Majorana fermions (across 3 generations) for a choice of S and Ω.

Table 1.4.2   For non-zero-mass particles, D, the number of generations, and the number of particles
- D: (1.4.4)
  - Cases
    - D(non-zero-mass boson fields) = $3-\Omega$
    - D(non-zero-mass boson particles) = $3-\Omega$
    - D(fermion fields) = $(5-4\Omega)/2$
    - D(fermion particles) = $(21-4\Omega)/6$
  - For each case that correlates with elementary particles or their fields, D is a positive integer
- Number of generations correlating with each elementary fermion: (1.4.5)
  - 3
- Number of particles: (1.4.6)
  - For non-zero-mass bosons
    - 2S+1
  - For fermions, assuming none of the fermions is a Majorana fermion
    - 2(2S+1), for each of 3 generations
    - 6(2S+1), across 3 generations
  - For fermions, assuming each of a maximal number of the fermions is a Majorana fermion
    - D(fermion particles), for each of 3 generations
- AUL#MF (an upper limit on the number of Majorana fermions, across 3 generations):   (1.4.7)
  - 3 × (2(2S+1) − D(fermion particles))

~ ~ ~

The next table provides alternative designations for oscillators. We use the linear numbering designations to, for example, simplify the labelling of solutions.

Table 1.4.3   Two sets of names for oscillators

E	E	E	E	E	E	E	P	P	P	P	P	P	P	P	P	polarization-	(1.4.8)
6R	6L	4R	4L	2R	2L	0	0	2L	2R	4L	4R	6L	6R	8L	8R	centric	(1.4.9)
E	E	E	E	E	E	E	P	P	P	P	P	P	P	P	P	linear	(1.4.10)
6	5	4	3	2	1	0	0	1	2	3	4	5	6	7	8	numbering	(1.4.11)

Section 1.5   The W-family, H-family, and O-family

Abs.1.5.1   We list all known and some possible non-zero-mass elementary bosons.
- This work points to opportunities to discover or infer members of a family of non-zero-mass elementary bosons for which no particles have as yet been identified.

The next table lists solutions that might correlate with ground states for non-zero-mass elementary bosons. For each item in the table, Œ=0. Also, 2S=#P. Notation kW% correlates with the W-family. For the W-family, % correlates with QP-like oscillators. The symbol kH% that equals 0H0 correlates with the H-family. For the O-family, the % in kO% correlates with QE-like oscillators. Table 1.4.3 shows equivalences between 2 sets of names for oscillators. For each family, k=2S. The number of particles per kX% group (for X = W, H, or O) is 2k+1 = 2S+1, per item (1.4.6). People might correlate the term O-family with the letter o in leptoquark.

Table 1.5.1   Ground states for S≤1 non-zero-mass elementary bosons

E 6R	E 6L	E 4R	E 4L	E 2R	E 2L	E 0	P 0	P 2L	P 2R	P 4L	P 4R	P 6L	P 6R	P 8L	P 8R	Particle or Mode	
				@	@	@	0	#	#							2W0	(1.5.3)
				@	@	@	#	0	#							2W1	(1.5.4)
				@	@	@	#	#	0							2W2	(1.5.5)
						0	0									0H0	(1.5.6)
				0	#	#	@	@	@							202	(1.5.7)
				#	0	#	@	@	@							201	(1.5.8)
				#	#	0	@	@	@							200	(1.5.9)

(1.5.1)
(1.5.2)

We associate some math solutions with known particles. Here, Q' denotes charge, in units of $|q_e|$, in which q_e denotes the charge of an electron.

Table 1.5.2   W- and H-family particles

Particle	Symbol	Q'	
Z boson	2W0	0	(1.5.11)
W⁺ boson	2W1	+1	(1.5.12)
W⁻ boson	2W2	−1	(1.5.13)
Higgs boson	0H0	0	(1.5.14)

(1.5.10)

In Section 4.2, we calculate charges for O-family bosons. The next table summarizes some results.

Table 1.5.3   Charges for S=1 O-family bosons
- 202 bosons have Q' = −1/3   (1.5.15)
- 201 bosons have Q' = +1/3
- 200 bosons have Q' = 0

As with quarks, O-family bosons cannot be created or observed as free particles.

~ ~ ~

For an interaction vertex in which a W⁻ and a neutrino enter and an electron exits, the next table shows before and after states for the W boson. Here, we assume no other W-family bosons or excitements pertain.

Here, we ignore the QE-like symmetry. (We, address the topic of such symmetry in Section 3.4 and Section 5.1.)

Table 1.5.4 2W% states related to a specific interaction vertex

E	E	E	E	E	E	P	P	P	P	P	P	P	P	P	P	Particle or	
6R	6L	4R	4L	2R	2L	0	0	2L	2R	4L	4R	6L	6R	8L	8R	Mode	(1.5.16) (1.5.17)
				#	#	1	#	#	1							2W2 before	(1.5.18)
				#	#	0	#	#	0							2W2 after	(1.5.19)

Section 1.6   The Q-family and L-family

Abs.1.6.1   We list all known non-zero-mass elementary fermions.
- This work points to ways to represent quark interaction-vertices and lepton interaction-vertices via harmonic oscillator math.

The next table lists types of known members of the Q-family and the L-family. We use item (1.4.4) to compute values of D. For the notation 1L, the 1 denotes $2S=1$ and the L denotes L-family. For the notation 1Q, the 1 denotes $2S=1$ and the Q denotes Q-family. For the Q-family, for $S=1/2$, no possible other particles exist. We use item (1.4.5) to note that each particle has 3 generations. We use item (1.4.6) to compute, assuming all particles are Dirac particles, numbers of particles per generation and total number of particles. We use item (1.4.7) to compute AUL#MF, an upper limit on the number of Majorana fermions, across 3 generations.

Table 1.6.1   Types of S=1/2 L-family and Q-family particles

S	Ω	D for fermion fields	D for fermion particles	Known types of particles	Possible other particles	Number of particles (if all Dirac) • per gen • total	AUL#MF	(1.6.1)
1/2	3/4	1	3	Leptons (1L)		• 4 • 12	3	(1.6.2)
1/2	−3/4	4	4	Quarks (1Q)		• 4 • 12	0	(1.6.3)

In Part 3, we use Œ=0 math to show that solutions may correlate with interaction vertices in which elementary fermions participate. Here, we state some results.

Here, we focus on interactions that change basic properties of fermions. We use the term basic property so as to include charge. We do not include generation, spin-orientation, or momentum in this concept of basic property.

~ ~ ~

The next table points to possible interaction vertices for generation-1 quarks (or, generation-1 1Q particles). We show interactions for which 1 generation-1 spin-1/2 elementary fermion and 1 spin-1 non-zero-mass elementary boson enter and 1 spin-1/2 elementary fermion exits. In each interaction, the incoming fermion absorbs property from an O-family or W-family boson. Here, Q denotes any unit of charge. For instances of Q in this table, values (in units of $|q_e|$) of absorbed charge are −1/3 (for E2R), +1/3 (for E2L), +1 (for P2L), and −1 (for P2R). For example, a down quark can absorb +1 unit of positive charge from a W$^+$ and thereby become an up quark. Or, by absorbing +1/3 unit of charge from a 2O1, an anti-down becomes an up. For interactions between quarks and charged O-family members, parity violations occur.

Table 1.6.2   Property-absorption vertices for generation-1 spin-1/2 Q-family fermions

E 6R	E 6L	E 4R	E 4L	E 2R	E 2L	0	0	P 2L	P 2R	P 4L	P 4R	P 6L	P 6R	P 8L	P 8R	Particle or Mode	
																Particle or	(1.6.4)
																Mode	(1.6.5)
				Q					Q							up	(1.6.6)
					Q				Q							anti-down	(1.6.7)
				Q				Q								down	(1.6.8)
					Q			Q								anti-up	(1.6.9)

People might correlate each row in Table 1.6.2 with a W- and/or O-family destruction operator that pertains for the generation-1 quark the particle or mode column refers to.

The pattern in Table 1.6.2 also applies for generation-2 and generation-3 quarks.

~ ~ ~

Regarding Table 1.6.2, people might say the E0 and P0 oscillators are not active. People might say that the number of relevant QP-like oscillators is 2. For non-zero-mass bosons, having exactly 1 relevant QP-like oscillator correlates with 2S=1−1=0. For non-zero-mass bosons, having exactly 3 relevant QP-like oscillators correlates with 2S=3−1=2. As with representations for elementary bosons, the above representations regarding elementary fermions pertain to interactions. For the fermions in the above table, the next item pertains. For the purposes of computing fermion spin, people might say that the P0 oscillator is not relevant.

Gss.1.6.1   For elementary fermions, 2S=#P−1.   (1.6.10)

~ ~ ~

The next table provides a symbol for each of the 4 generation-1 quarks. For the symbol 1Q(j,k), 1 denotes 2S. The symbol j correlates with the 1 active QE-like oscillator, out of the pair E2R and E2L. The symbol k correlates with the 1 active QP-like oscillator out of the pair P(#P)L and P(#P)R. Here, #P=2.

Table 1.6.3   Symbols for generation-1 quarks

Quark	Symbol	
		(1.6.11)
up	1Q(2R,2R)	(1.6.12)
anti-down	1Q(2L,2R)	(1.6.13)
down	1Q(2R,2L)	(1.6.14)
anti-up	1Q(2L,2L)	(1.6.15)

In item (3.4.209), we note that all 3 generations of a fermion correlate with 1 field. For the purposes of this paper, we see no significant harm in not providing tables or symbols for generation-2 or generation-3 Dirac elementary fermions. For the rest of this section, we downplay the concept of generation.

The next table summarizes some interactions for S=1/2 Q-family elementary fermions. The table echoes results noted above. The symbol 1Q correlates with S=1/2. For each symbol for a particle, the first item in parenthesis correlates with the QE-like oscillator for which N(...)=−2. For each symbol for a particle, the second item in parenthesis correlates with the QP-like oscillator for which N(...)=−2. The table summarizes Q-family interactions with some O- and W-family bosons. We show interactions with spin-1 O-family bosons (that is, 2O bosons) and W-family bosons (that is, 2W bosons). Each item in the next table satisfies

$Œ=0$. The symbol −1 indicates that no property-changing interaction can occur. The symbol −2 indicates that an interaction-vertex exists.

Table 1.6.4   Property-absorption vertices for S=1/2 Q-family elementary fermions   (1.6.16)

E 6R	E 6L	E 4R	E 4L	E 2R	E 2L	P 0	P 0	P 2L	P 2R	P 4L	P 4R	P 6L	P 6R	P 8L	P 8R	Particle or Mode	(1.6.17)
				−2	−1	−1	−1	−1	−2							1Q(2R,2R)	(1.6.18)
				−1	−2	−1	−1	−1	−2							1Q(2L,2R)	(1.6.19)
				−2	−1	−1	−1	−2	−1							1Q(2R,2L)	(1.6.20)
				−1	−2	−1	−1	−2	−1							1Q(2L,2L)	(1.6.21)

Table 1.6.4 dovetails with Table 1.6.1 regarding numbers of fermions.

~ ~ ~

The next table points to property-changing interaction vertices for generation-1 leptons. We show interactions for which 1 generation-1 spin-1/2 elementary fermion and 1 spin-1 non-zero-mass boson enter and 1 spin-1/2 elementary fermion exits. We list 2 neutrinos, e-neutrino (electron-neutrino) and p-neutrino (positron-neutrino). People might say that listing 2 neutrinos corresponds to assuming Dirac-fermion pertains. In each interaction, the incoming fermion absorbs property from a W-family boson. Here, Q denotes any unit of charge. For instances of Q in this table, values of absorbed charge include +1 (for P2L) and −1 (for P2R). For example, an electron can absorb a unit of +1 charge from a W⁺ but cannot absorb not a unit of −1 charge from a W⁻. By absorbing +1 unit of positive charge, the electron becomes a neutrino.

Table 1.6.5   Property-changing vertices for generation-1 L-family fermions (assuming Dirac neutrinos)   (1.6.22)

E 6R	E 6L	E 4R	E 4L	E 2R	E 2L	P 0	P 0	P 2L	P 2R	P 4L	P 4R	P 6L	P 6R	P 8L	P 8R	Particle or Mode	(1.6.23)
								Q								electron	(1.6.24)
									Q							positron	(1.6.25)
									Q							e-neutrino	(1.6.26)
								Q								p-neutrino	(1.6.27)

Table 1.6.5 dovetails with Table 1.6.1 regarding numbers of Dirac fermions.

For purposes of providing symbols, we continue to assume (for the moment) that neutrinos are Dirac fermions. The next table provides a symbol for each of 4 generation-1 leptons. For the symbol 1L(j,k), 1 denotes 2S. The symbol j correlates with charge. The notation j = Q'≠0 correlates with the particle's having non-zero charge. The notation j = Q'=0 correlates with the particle's having 0 charge. The symbol k correlates with the 1 active QP-like oscillator out of the pair P2L and P2R.

Table 1.6.6   Symbols for generation-1 leptons, assuming neutrinos are Dirac fermions

Lepton	Symbol		(1.6.28)
electron	1L(Q'≠0;2L)		(1.6.29)
positron	1L(Q'≠0;2R)		(1.6.30)
e-neutrino	1L(Q'=0;2R)		(1.6.31)
p-neutrino	1L(Q'=0;2L)		(1.6.32)

The next table pertains to each of the 3 generations of the L-family. For each item row, $Œ=0$.

Table 1.6.7   Property-absorption vertices for L-family elementary fermions (assuming Dirac neutrinos)

E	E	E	E	E	E	E	P	P	P	P	P	P	P	P	P	Particle or	
6R	6L	4R	4L	2R	2L	0	0	2L	2R	4L	4R	6L	6R	8L	8R	Mode	(1.6.33)(1.6.34)
		−1	−1	−1	−1	−1	−1	−2	−1							1L(Q'≠0;2L)	(1.6.35)
		−1	−1	−1	−1	−1	−1	−1	−2							1L(Q'≠0;2R)	(1.6.36)
		−1	−1	−1	−1	−1	−1	−1	−2							1L(Q'=0;2R)	(1.6.37)
		−1	−1	−1	−1	−1	−1	−2	−1							1L(Q'=0;2L)	(1.6.38)

To the extent neutrinos are Majorana fermions, people might say that $2^{-1/2}(\langle\text{e-neutrino}| + \langle\text{p-neutrino}|)$ correlates with the destruction operator relevant to interactions with the W- and O-families. An equivalent expression is $2^{-1/2}(\langle 1L(Q'=0,2R)| + \langle 1L(Q'=0,2L)|)$. Symbols for creation operators would be similar, but with subtractions in place of additions.

The next item pertains to the extent neutrinos are Majorana fermions. For each row, Œ=0.

Table 1.6.8   Property-absorption vertices for L-family elementary fermions (assuming Majorana neutrinos)

E	E	E	E	E	E	E	P	P	P	P	P	P	P	P	P	Particle or	
6R	6L	4R	4L	2R	2L	0	0	2L	2R	4L	4R	6L	6R	8L	8R	Mode	(1.6.39)(1.6.40)
		−1	−1	−1	−1	−1	−1	−2	−1							1L(Q'≠0;2L)	(1.6.41)
		−1	−1	−1	−1	−1	−1	−1	−2							1L(Q'≠0;2R)	(1.6.42)
−1	−1	−1	−1	−1	−1	−1	−1	−2	−2							1L(Q'=0;2L, 2R)	(1.6.43)

For 1L particles, no interactions occur with O-family bosons (or with Y-family bosons). The next table applies.

Table 1.6.9   Handedness for leptons

- For a 1L(Q'≠0;2j) lepton …   (1.6.44)
  - People might say that j=L correlates with the term left-handed
  - People might say that j=R correlates with the term right-handed
- To the extent neutrinos are Dirac fermions, for a 1L(Q'=0;2j) lepton …   (1.6.45)
  - People might say that j=L correlates with the term right-handed
  - People might say that j=R correlates with the term left-handed
- To the extent neutrinos are Majorana fermions, …   (1.6.46)
  - People might say that no handedness pertains

Section 1.7   The Y-family

Abs.1.7.1   The Y-family provides a basis for gluons.
- This work provides insight about gluons and related symmetries.

The Y-family provides a basis for gluons.

For the G-family, people might say that N(P0)<0 correlates with a lack of longitudinal excitation. For the Y-family, the next table provides a representation in which no QP-like excitation takes place with respect to fermions. We continue to use Œ=0. For the notation 2Yjk, the 2 denotes that the spin S correlates with 2S=#P=2. The j denotes the Ej for which N(Ej)=−2. (Regarding the notation Ej, see Table 1.4.3.) The k denotes the Ek for which N(Ek)=0.

Table 1.7.1   Y-family states of a 1Q fermion

E	E	E	E	E	E	E	P	P	P	P	P	P	P	P	P	Particle or	
6R	6L	4R	4L	2R	2L	0	0	2L	2R	4L	4R	6L	6R	8L	8R	Mode	(1.7.2)
				−2	−1	0	−1	−1	−1							2Y20	(1.7.3)
				0	−2	−1	−1	−1	−1							2Y12	(1.7.4)
				−1	0	−2	−1	−1	−1							2Y01	(1.7.5)
				−1	−2	0	−1	−1	−1							2Y10	(1.7.6)
				0	−1	−2	−1	−1	−1							2Y02	(1.7.7)
				−2	0	−1	−1	−1	−1							2Y21	(1.7.8)

(1.7.1)

The above representations are consistent with item (1.3.10).
To describe fields for Y-family members, we add 1 to each N in Table 1.7.1.

Table 1.7.2   Ground states for spin-1 Y-family particles and fields

E	E	E	E	E	E	E	P	P	P	P	P	P	P	P	P	Particle or	
6R	6L	4R	4L	2R	2L	0	0	2L	2R	4L	4R	6L	6R	8L	8R	Mode	(1.7.10)
				−1	0	1	@	@	@							2Y20	(1.7.11)
				1	−1	0	@	@	@							2Y12	(1.7.12)
				0	1	−1	@	@	@							2Y01	(1.7.13)
				0	−1	1	@	@	@							2Y10	(1.7.14)
				1	0	−1	@	@	@							2Y02	(1.7.15)
				−1	1	0	@	@	@							2Y21	(1.7.16)

(1.7.9)

For each Y-family elementary boson, the next items pertain. Here, the symbols CCj denote color charges. The color charges relevant to 2Y bosons are CC2, CC1, and CC0 (as in E2R, E2L, and E0, respectively). Here, the values of N(Ej) and N(Ek) correlate with Table 1.7.2 (and not with Table 1.7.1).

- For exactly 1 value of j, N(Ej)=−1   (1.7.17)
- For exactly 1 value of k, N(Ek)=1   (1.7.18)
- An excitement includes the following   (1.7.19)
  - A quark loses color charge CCj
  - N(Ej) becomes 0
  - N(Ek) becomes 0

People might say that such an excitement erases color charge CCj from a quark.
Similarly, a de-excitement from a state having N(E2R) = N(E2L) = N(E0) = 0 paints the quark with a color charge.
In the next item, 1 trio consists of items (1.7.11), (1.7.12), and (1.7.13). Here, the cyclic order for N(El) values is −1, 0, 1. We call this the ascending-order trio. Another trio consists of items (1.7.14), (1.7.15), and (1.7.16). Here, the cyclic order for N(El) values is 1, 0, −1. We call this the descending-order trio.

> Gss.1.7.1   One trio of Y-family bosons provides for gluons pertaining to quarks (1.7.20)
> people consider to be matter. The other trio pertains to quarks people
> consider to be antimatter.

For discussion, we assume that the first trio corresponds to quarks people consider to be matter (and that the second trio corresponds to quarks people consider to be antimatter). (Possibly, the reversed pairing pertains.) The next item symbolizes a component for a gluon. The right element erases color charge CC2 from a quark. The left element paints color charge CC3.

$$| \text{item } (1.7.11) > < \text{item } (1.7.12) | \qquad\qquad (1.7.21)$$

People sometimes denote the 3 color charges by r (for red), b (for blue), and g (for green). For such, we use l' to denote erasing color charge l. We use n to denote painting color charge n. For discussion, we assume CC1 corresponds to r and CC2 corresponds to b. The next item restates item (1.7.21).

$$\text{br'} \qquad\qquad (1.7.22)$$

The next items show 2 gluons of which item (1.7.22) comprises a component.

$$(rb' + br') / 2^{1/2} \qquad\qquad (1.7.23)$$
$$-i(rb' - br') / 2^{1/2} \qquad\qquad (1.7.24)$$

The next item provides a way people symbolize another 1 of the 8 gluons.

$$(rr' + bb' - 2gg') / 6^{1/2} \qquad\qquad (1.7.25)$$

The next item pertains.

- g (in the r, b, g representation for color charge) correlates with CC0 (in the CC2, CC1, and CC0 representation)   (1.7.26)

People can consider that $\Omega = -S(S+1) = -2$ for spin-1 members of the Y-family.

People can correlate the number, 8, of gluons (correlating with either matter quarks or antimatter quarks) with the number of generators for the group SU(3). Here, the 3 in SU(3) correlates with the count, 3, of QE-like oscillators.

People can correlate the group U(1) with the relevance of the 2 trios - the ascending-order trio and the descending-order trio. The group SU(2) has 3 generators. People can correlate SU(2) with the 3 Y-family members in a trio. People might correlate these results with the term QE-like. As yet, we have not considered how to correlate our representations of interactions with QP-like space-time coordinates. That correlation correlates with an SU(3) symmetry. (There are 3 QP-like coordinates.) Thus, the symmetry SU(3)×SU(2)×U(1) correlates with the spin-1 part of the Y-family. People correlate SU(3)×SU(2)×U(1) symmetry with the Standard Model.

Part 2   Models that correlate with phenomena

Section 2.0   Introduction and summary

Abs.2.0.1   We correlate observed phenomena with above-mentioned elementary particles.
- This work provides opportunities to resolve various physics-theory problems people associate with the term unsolved.

In Part 2, we suggest ways to resolve problems that people associate with the concept that traditional theoretical physics does not adequately correlate with observed phenomena. People might associate the problems with the areas of elementary-particle physics, astrophysics, and cosmology. Topics include the rate of expansion of the universe, curvature (if any) of the universe, baryon asymmetry, violations of CPT-related symmetries, quasars, and the number (3) of generations of fermions. Also, we predict a tauon mass $(1.776814(\sim48)\times10^3$ MeV/c^2) with a smaller uncertainty range than experiments have yielded.

Section 2.1   Long-term changes in the rate of expansion of the universe

Abs.2.1.1   The G-family particles 84G2468&, 64G246&, and 42G24& provide for changes in the rate of expansion of the universe.
- This work suggests that forces (and not just the existence and some effects of dark-energy stuff) regulate the observed expansion of the universe.

The next items pertain. For a jkG%& boson, −N(P0) correlates with the number of elements in %. We base item (2.1.1) on the possibility that a factor of r^{-2} correlates with each element. (We use the word approximates because, for computations simulating large objects, people may need to consider differences {across 1 or more objects} in the times at which relevant G-family bosons are emitted.)

> Gss.2.1.1   The spatial dependence of the force associated with a G-family   (2.1.1)
> boson approximates $r^{2\times N(P0)}$. Here, r denotes the distance between the
> centers of property (such as charge {for 22G2&, the electromagnetic force}
> or mass or mass-energy {for 44G4&, gravity}). Here, N(P0) pertains to the
> G-family boson.
> - For example, 2×N(P0) =   (2.1.2)
>   - −2 for 22G2& (photons) and for 44G4& (gravitons)
>   - −4 for 42G24&
>   - −6 for 64G246&
>   - −8 for 84G2468&

We think about 2 clumps of stuff. These clumps have similar size and similar amounts of material. The clumps neighbor each other.

The next items discuss forces that dominate within each clump and between the 2 clumps. These items reflect the observation that the universe expands. During such expansion, effects of an r^{-2j} force (within or between the 2 clumps) evolve from being more than effects of an $r^{-2(j-1)}$ force to being less than effects of an $r^{-2(j-1)}$ force. The next items define 4 eras.

- During era FE1, the force associated with 84G2468& dominates   (2.1.3)
- During era FE2, the force associated with 64G246& dominates   (2.1.4)
- During era FE3, the force associated with 42G24& dominates   (2.1.5)
- During era FE4, the forces associated with 44G4& and 22G2& dominate   (2.1.6)

Today, observed atoms, planets, stars, galaxies, and galactic clusters exhibit era FE4 behavior. At and beyond a size that is larger than sizes people associate with galactic superclusters, people observe FE3 behavior.

People deduce rates of expansion of the universe from observations of photons. People express approximate time after the big bang in terms of the redshift, z, people correlate with photons that people and equipment observe. Redshift $z=0$ pertains to photons emitted recently. Redshift numbers are positive. A value of z denotes a time before the time people associate with a smaller value of z.

The next table correlates some redshifts with times after the big bang. [Ref.2.1.1]

Table 2.1.1   Some redshifts and the related times after the big bang

z	Time after the big bang (years)	
		(2.1.7)
2.3	2.9×10^9	(2.1.8)
0.7	7.3×10^9	(2.1.9)
0.46	8.9×10^9	(2.1.10)
0	13.8×10^9	(2.1.11)

The next items pertain. [Ref.2.1.2 and Ref.2.1.3] 3 eras exist.

- For $z >$ some number greater than 2.3, the observed rate of expansion   (2.1.1)
  increases
  - We call this era ZE1
- For $\sim 2.3 > z > \sim 0.7$, the observed rate of expansion decreases   (2.1.2)
  - We call this era ZE2
- For $0.46 \pm 0.13 > z > 0$, the observed rate of expansion increases   (2.1.3)
  - We call this era ZE3

The next items pertain.

Gss.2.1.2   For observed astrophysical objects of above some size, era FEk   (2.1.4)
correlates with era ZEn, for $1 \leq k = n \leq 3$.

Gss.2.1.3   84G2468& repels astrophysical objects from each other. 64G246&   (2.1.5)
attracts astrophysical objects to each other. 42G24& repels astrophysical
objects from each other.

~ ~ ~

Ref.2.1.1   N. Gnedin, Cosmological Calculator for the Flat Universe. (http://home.fnal.gov/~gnedin/cc/)
Ref.2.1.2   N. G. Busca, et. al., Baryon Oscillations in the Lyα forest of BOSS quasars, arXiv:1211.2616 [astro-ph.CO].
Ref.2.1.3   A. Riess, et. al., Type Ia Supernova Discoveries at z > 1 from the *Hubble Space Telescope*: Evidence for Past Deceleration and Constraints on Dark Energy Evolution, *The Astrophysical*

Journal, 607, 665 (2004). (doi:10.1086/383612) (http://iopscience.iop.org/0004-637X/607/2/665)

Section 2.2   Curvature of the universe

Abs.2.2.1   Dominance of the G-family forces 84G2468&, 64G246&, and 42G24& correlates with the universe's having little or no large-scale curvature.

- This work suggests that forces and symmetries (and not a critical density) correlate with near or actual flatness of the universe.

People discuss the extent to which to consider that, on large scales, the universe exhibits positive curvature (or, is spherical or has $\Omega_0 > 1$), no curvature (or, is flat or has $\Omega_0 = 1$), or negative curvature (or, is hyperbolic or has $\Omega_0 < 1$). (Ω_0 is not related to the Ω or to the Ω_2 we use in this paper.) Observations indicate that $\Omega_0 \approx 1$ may pertain. [Ref.2.2.1] People associate the notion of a critical density with $\Omega_0 = 1$.

In Section 1.3, we discuss the 4 G-family #E=0 bosons 84G2468&, 64G246&, 42G24&, and 22G2&. We call these the G-family #E=0 series. This series correlates with 1 QE-like space-time coordinate and 3 QP-like space-time coordinates. Traditional astrophysics includes the concept of co-moving coordinates. People correlate co-moving coordinates with 1 QE-like dimension and 3 QP-like dimensions. The next items pertain. (We discuss inflation in Section 7.1.)

> Gss.2.2.1   Stuff that constitutes the currently observable universe exited the period of inflation in a state such that people can consider that any curvature was adequately small that people can consider that zero- or negligible-curvature pertains at the time of exit or somewhat thereafter.   (2.2.1)
>
> - To the extent 84G2468&, 64G246&, and 42G24& bosons dominated (after that exit or a somewhat later time) interactions between objects, people can consider those objects to be part of a universe for which near-zero curvature pertains   (2.2.2)
> - For measurements of era FE1, FE2, or FE3 phenomena, the dominance of no more than 84G2468&, 64G246&, and 42G24& forces correlates with the observation that $\Omega_0 \approx 1$ (2.2.3)

The next items correlate with work above. We note regarding 44G4& that #E=2.

> - For era FE1, FE2, or FE3 phenomena, people can use a space-time coordinate system that correlates with little or no curvature   (2.2.4)
> - For observations expressed in terms of co-moving coordinates, people might say that the force 44G4& (gravity) contributes to a need to use space-time coordinate systems for which people associate non-zero curvature   (2.2.5)

~ ~ ~

Ref.2.2.1   NASA, http://map.gsfc.nasa.gov/universe/uni_shape.html

Section 2.3   Baryon asymmetry (or, matter/antimatter imbalance)

Abs.2.3.1   Spin-1 O-family bosons could have catalyzed the current matter/antimatter imbalance.
-    This work provides a possible basis for resolving the problem of baryon asymmetry.

Perhaps, for some era in the past, the ratio of 1L-plus-1Q matter to 1L-plus-1Q antimatter was less than it is now. Perhaps such an era occurred before stuff became sufficiently dispersed that 2O2 and 2O1 bosons would have limited impact.

To the extent such an era featuring more equal matter/antimatter balance occurred, we suggest a mechanism that could have changed the ratio.

Interactions between quarks and 20% bosons (for % = 2 or 1) convert antimatter quarks into matter quarks and vice versa. Lasing of 20% bosons could have led to, for example, a transformation from equal amounts of matter and antimatter to the imbalance people correlate with today's universe. Presumably, even a small asymmetry regarding direction (antimatter quarks to matter quarks or vice versa) could have grown. Reactions that converted antimatter charged leptons (such as positrons) into neutrinos and then into matter charged leptons (such as electrons) could have occurred concurrently with the interactions that led to changes in ratios between matter quarks and antimatter quarks.

Section 2.4   Violations of CPT-related symmetries

Abs.2.4.1   O-family bosons correlate with CPT-related symmetry violations.
-    This work points to how to resolve problems related to the sizes of CPT-related symmetry violations exceeding symmetry-violation sizes people correlate with Standard Model physics.

The next items discuss traditional thinking regarding violations of CPT-related symmetries.

- The Standard Model embraces interactions mediated by spin-0 and spin-1 elementary bosons   (2.4.1)
- The Standard Model does not include the O-family   (2.4.2)
- People discuss at least 3 types of CPT-related symmetry violations ...   (2.4.3)
  - P-violations (parity violations)
  - C-violations (charge conjugation violations)
  - CP-violations
- People say that the Standard Model correlates with some instances of the first 2 types of violations   (2.4.4)
- People say that, for each of the 3 types of violations, the Standard Model does not correlate with magnitudes of violations that are as big as observed magnitudes   (2.4.5)

The next table pertains. The term interact indirectly with contrasts with the term interact directly with. The term interacts directly with denotes has an interaction vertex with.

Table 2.4.1   Possible sources of CPT-related violations, beyond Standard Model violations

- Charged O-family bosons transform antimatter quarks into matter quarks   (2.4.6)
  or vice versa
  - Such an interaction would involve each of 2 vertices
  - People might say that each vertex correlates with P-violation for a
    fermion
- Baryonic matter fermions interact indirectly with Majorana neutrinos,   (2.4.7)
  sibling-state zero-charge fermions (See, for example, Section 3.4 and Part
  5.), or other dark-matter fermions

Section 2.5   Quasars

Abs.2.5.1   At least 1 of the forces 42G24& and 84G2468& produces quasars.
- The work provides possible mechanisms that lead to formation of quasars.

People correlate each observed quasar with a black hole. For each of many black holes, observations do not include a quasar. The next items pertain.

- While and after a black hole first forms, the density of matter increases   (2.5.1)
  based on the influence of 44G4&
- Eventually, the density may become large enough that 42G24& and/or   (2.5.2)
  84G2468& repulsion dominates 44G4& attraction
  - Collapse slows
  - Collapse may partly reverse
    - A quasar may develop

Section 2.6   Tauon mass

Abs.2.6.1   The mass of a tauon may be $1.776814(\sim48)\times10^3$ MeV/c^2.
- This work provides impetus to improve the accuracy of measurements of the mass of tauons
  and the accuracy of measurements of the gravitational constant.

The next item restates results from Section 3.5. Here, m_{tauon} denotes the mass of a tauon, q_e denotes the charge of an electron, $1/(4\pi\varepsilon_0)$ denotes the Coulomb constant, G_N denotes the gravitational constant, and m_e denotes the mass of an electron.

$$m_{tauon}/m_e = \beta' = \beta = \exp(\ (1/12)\ \log\{\ \{3/4\}\ \{(q_e)^2/(4\pi\varepsilon_0)\}\ /\ \{G_N(m_e)^2\}\ \})\qquad(2.6.1)$$

The next item predicts a mass for tauons. We base this prediction on items (3.5.2), (3.5.3), (3.5.4), and (3.5.6). The largest contributor to the uncertainty range item (2.6.2) shows comes from measurements of G_N. The uncertainty range item (2.6.2) shows may be an over-estimate.

$$m_{tauon} \approx 1.776814(\sim48)\times10^3\ MeV/c^2\qquad(2.6.2)$$

The next item shows experimental results. [Section 10.5]

$$m_{tauon} \approx 1.77682(16) \times 10^3 \text{ MeV}/c^2 \qquad (2.6.3)$$

The next items pertain.

Tbd.2.6.1   Verify or rule out (to some confidence level) the value for m_{tauon} we   (2.6.4)
predict.

Tbd.2.6.2   Verify or rule out (to some confidence level) that $\beta = \beta'$.   (2.6.5)

Section 2.7   Number of generations for fermions

Abs.2.7.1   Each elementary fermion is part of a 3-generation (or 3-flavor) trio of elementary fermions.
- This work solves a problem regarding providing a math model that correlates with fermions having 3 generations.

Item (3.2.59) correlates with each elementary fermion. For charged fermions, each elementary fermion correlates with a trio of elementary fermions. For charged fermions, people use the term 3 generations. For neutrinos, people may prefer to say that 3 flavors pertain. [Discussion near Table 4.4.6]

Part 3   Mathematics

Section 3.0   Introduction and summary

Abs.3.0.1   We discuss mathematics underlying much of this paper.
  ▪ This work opens (and provides applications of) a branch of math within isotropic quantum harmonic oscillators.

In Part 3, we provide mathematics that underlies much of this paper.

Section 3.1   Mathematical models for quantum phenomena

Abs.3.1.1   Physics observations include quantized results with which action/Lagrangian math models do not correlate.
  ▪ This work provides impetus to develop (from quantum-math bases that do not start from classical-physics action/Lagrangian math) inherently quantum approaches for modeling physics phenomena.

The next table features sets of discreet numbers.

Table 3.1.1   Types of observations that yield discreet numbers
  • For each known elementary particle, the expression $j|q_e|/3$, with j being an (3.1.1) integer, describes the charge of the particle
    • q_e denotes the charge of an electron
  • For each elementary particle, the expression $S\hbar$, with 2S being a non- (3.1.2) negative integer, describes the spin of the particle
    • $\hbar$ denotes Planck's constant (reduced)
  • Harmonic-oscillator-math raising operators and lowering operators (3.1.3) correlate with quantum descriptions of modes of the vector potential
    • For each photon, the spectrum of excitation states is discreet
  • The number $S(S+D_P-2)$, with $D_P=3$, pertains for the radial-component of (3.1.4) some harmonic-oscillator math correlating with particles having spin/$\hbar$ of S
    • $S(S+D_P-2) = S(S+1)$, for $D_P=3$
    • For each known elementary particle, people consider such an $S(S+1)$ to be relevant to the physics of the particle
  • For each elementary particle, a non-negative number m, describes the mass (3.1.5) of the particle
    • The spectrum of such masses is discreet
  • The single speed c and the single expression $E^2-c^2P^2=m^2c^4$ pertain for all (3.1.6) free elementary particles
    • In the expression, E denotes energy, c denotes the speed of light (in a vacuum), P denotes momentum, and m denotes the mass of the specific particle
    • Free particles include electrons and photons
    • Free particles do not include quarks and gluons

- For each known elementary fermion, the number of generations is 3   (3.1.7)

People try to develop mathematics that has solutions that people can correlate with observations about nature. Some such attempts have bases in the concept of action and use Lagrangian math. Results include the Standard Model. People correlate the Standard Model with some observations about elementary particles, compound particles, and interactions between particles. People say that the Standard Model falls short regarding various observations.

People have tried to use action/Lagrangian approaches to unify gravitation and quantum mechanics. People have had access to general relativity since 1916. People have had access to aspects of quantum mechanics since about the same time.

People might say that action/Lagrangian approaches have yet to cover the entirety of the topics to which Table 3.1.1 alludes. People might say that such approaches do not predict masses for some elementary particles. People might say that action/Lagrangian approaches have yet to adequately unify gravitation and quantum mechanics.

We think that exploring math other than action/Lagrangian math may be useful. We think that trying to catalog elementary particles and their interactions may be useful. We think that starting from a quantum-math basis may be useful.

~ ~ ~

People might say that item (3.1.3) pertains exactly, at least to the extent of known photonics.

People might say that other uses of harmonic-oscillator math in traditional physics represent attempts to layer quantum math on top of math for non-quantum physics. People might say that such attempts feature approximations to physics phenomena. For example, attempts to quantize aspects of classical physics exactly via harmonic-oscillator math would require modelling an infinitely large potential energy.

~ ~ ~

Harmonic-oscillator math correlates with at least 1 seemingly exact model for quantum phenomena. We base our work on mathematics for harmonic oscillators.

Section 3.2   Mathematics for isotropic quantum harmonic oscillators

Abs.3.2.1   We discuss mathematics for isotropic quantum harmonic oscillators.
- This work shows new solutions within math related to isotropic quantum harmonic oscillators.

We denote 2 numbers. We correlate #E with the term QE-like. We correlate #P with the term QP-like.

- #E denotes a non-negative integer   (3.2.1)
- #P denotes a non-negative integer   (3.2.2)

The next items denote 2 sets. Each set consists of indices associated with a sequence of consecutive non-negative integers.

$$\text{SIDE} = \{\, E_j \mid j = \#E\,,\ \#E{-}1\,,\ \dots\,,\ 1\,,\ \text{or } 0\,\}$$ (3.2.3)
$$\text{SIDP} = \{\, P_j \mid j = 0\,,\ 1\,,\ \dots\,,\ \#P{-}1\,,\ \text{or } \#P\,\}$$ (3.2.4)

The next item denotes the union of the 2 sets.

$$SID = SIDE \cup SIDP \tag{3.2.5}$$

We use math for which the next items pertain. Here, $\in$ denotes belongs to (or, is a member of). Here, $N(j)$ denotes the quantum number for harmonic oscillator j. Each $N(j) + 1/2$ term correlates with a result from traditional quantum harmonic-oscillator math. Regarding item (3.2.8), the term isotropic correlates with each $\pm_j$ having the same magnitude as each other $\pm_j$. The difference in signs (+1 vis-à-vis −1) has significance. The reverse choice of signs can also work.

- #E is even and #P is even $\tag{3.2.6}$

$$0 = Œ = \Sigma_{j \in SID} \pm_j (N(j) + 1/2) \tag{3.2.7}$$
$$\pm_j = +1 \text{ for } j \in SIDE \tag{3.2.8}$$
$$\pm_j = -1 \text{ for } j \in SIDP$$

The next item repeats a feature of item (3.2.7).

$$Œ = 0 \tag{3.2.9}$$

~ ~ ~

The next items pertain for a traditional-math ground state of an isotropic harmonic oscillator with 3 spatial (or, QP-like) dimensions and 1 temporal (or, QE-like) dimension. Item (3.2.12) contributes +3/2 to item (3.2.7). Item (3.2.13) contributes −3/2 to item (3.2.7). People would correlate the energy of this ground state with $N(E0) + 1/2 = 3/2$.

$$\#E = 0 \tag{3.2.10}$$
$$\#P = 2 \tag{3.2.11}$$
$$N(E0) = 1 \tag{3.2.12}$$
$$N(P0) = N(P1) = N(P2) = 0 \tag{3.2.13}$$

Table 1.1.3 exhibits 2 solutions for which we say the energy correlates with $N(E0) = 0 + 1/2 = 1/2$. There, as elsewhere in this paper, we use the notation Table 1.4.3 shows. People might use the term non-traditional (or the term new) to characterize these solutions.

~ ~ ~

We explore math for QP-like isotropic harmonic oscillators. (Regarding this work, people might say that we emphasize SIDP and downplay SIDE. [Item (3.2.5)])

The math pertains to solutions we denote by $\Psi(r)$. Here, r is the radial coordinate. Ψ can be a function also of coordinates (that is, angular coordinates) other than r. People call item (3.2.15) the Laplacian operator for D dimensions. People call item (3.2.22) the potential.

$$\xi \Psi(r) = (\xi'/2)(-\eta^2 \nabla^2 + \eta^{-2} r^2) \Psi(r) \tag{3.2.14}$$
$$\nabla^2 = r^{-(D-1)}(\partial/\partial r)(r^{D-1})(\partial/\partial r) - \Omega r^{-2} \tag{3.2.15}$$
- ξ and $\xi'/2$ denote numbers $\tag{3.2.16}$
- $\Psi(r)$ denotes a wave function $\tag{3.2.17}$
- r denotes a variable, with dimensions of length $\tag{3.2.18}$
- η denotes a length $\tag{3.2.19}$
- Ω denotes a number $\tag{3.2.20}$

- D denotes a positive integer   (3.2.21)
$$V = (\xi'/2)\, \eta^{-2}\, r^2 \qquad (3.2.22)$$

The next item pertains.

- For D=1, some solutions feature the following   (3.2.23)
  - $\Omega = 0$
  - The range $-\infty < r < \infty$ pertains
  - Ψ has the form of a Hermite polynomial (in the variable r) multiplied by $\exp(-r^{-2}/(2\eta^2))$

Work below tends not to use solutions people associate with (3.2.23). Below, the range $0 \leq r < \infty$ pertains for traditional solutions.

The next item describes solutions other than solutions people traditionally associate with (3.2.23).

$$\Psi(r) \propto r^{\nu} \exp(\, -r^{-2} / (2\eta^2)\,) \qquad (3.2.24)$$

The next table pertains for solutions for which item (3.2.23) does not pertain. The parameter η does not appear in these items.

Table 3.2.1   Algebraic relations pertaining to some solutions for isotropic harmonic-oscillator math
$$\xi = (D+2\nu)\, (\xi'/2) \qquad (3.2.25)$$
$$\Omega = \nu(\nu+D-2) \qquad (3.2.26)$$

The next items pertain to traditional solutions.

- ν is non-negative   (3.2.27)
- ν is an integer   (3.2.28)
- Ω is non-negative   (3.2.29)

Each of the next items points to non-traditional solutions.

- ν can be negative   (3.2.30)
- ν can be other than an integer   (3.2.31)

For D>2, item (3.2.30) is necessary (but not sufficient) for the next item to pertain.

- Ω can be negative   (3.2.32)

We limit our attention to solutions that can be normalized.
The next item shows behavior of the r-related normalization integrand near r=0.

$$\Psi^*\Psi\, r^{D-1} \sim r^{D-1+2\nu} \exp(-2r^2 2^{-1}\eta^{-2}) \sim r^{D-1+2\nu}, \text{ for } r\sim 0 \qquad (3.2.33)$$

The next items pertain to some solutions that normalize.

$$-1 < D-1+2\nu \qquad (3.2.34)$$
$$-D/2 < \nu \qquad (3.2.35)$$
- Ψ normalizes if (but not only if) ... $-(D/2) < \nu$ (3.2.36)

Thomas.J.Buckholtz@gmail.com   Copyright (c) 2014 Thomas J. Buckholtz   http://ThomasJBuckholtz.wordpress.com

The next item provides a definition of the Dirac delta function. [Ref.3.2.1]

$$\delta(r) = \lim_{\varepsilon \to 0+} (1/(2(\pi\varepsilon)^{1/2})) \exp(-r^2/(4\varepsilon)) \qquad (3.2.37)$$

We make the following association.

$$4\varepsilon = \eta^2 \qquad (3.2.38)$$

We assume that use of items (3.2.37) and (3.2.38) correlates with extending the range of integration. The next item shows an extended range of integration. Perhaps, people should consider that $r_E = \infty$.

$$-r_E \le r < \infty \qquad (3.2.39)$$
$$r_E > 0$$

For $r<0$, we note the possibility that the angular dependence of Ψ changes. For example, for θ being an angular coordinate, $\cos(\theta)$ for $r>0$ might become $\cosh(\theta)$ for $r<0$. We anticipate the possibility of products of exponentials and trigonometric functions.

The next item supplements item (3.2.36).

- Ψ normalizes if (but not only if) ... $-(D/2) = \nu$ (3.2.40)

We coin the next terms.

- Inside denotes $-(D/2) < \nu$ (3.2.41)
- Edge denotes $-(D/2) = \nu$ (3.2.42)

For each of inside or edge, η can have any real value other than 0. Two sets of mathematical solutions exist. One set corresponds to $\eta>0$. The other set corresponds to $\eta<0$.

For an edge case with -2ν an even integer, for each solution set, possibly 2 solutions exist. (For example, for $\Omega=0$, one candidate solution has $\Psi(-r)=\Psi(r)$. Another candidate solution has $\Psi(-r)=-\Psi(r)$. Here, because of items (3.2.37) and (3.2.39), the first candidate normalizes and the second candidate does not normalize.)

We call a linear combination (of candidate solutions) that normalizes a type-1 solution. We call a linear combination that does not normalize a type-2 solution.

We base the next item on considerations related to item (3.2.39).

       Gss.3.2.1   For an edge case with -2ν an even positive integer, 1 type-1 solution (3.2.43) exists.

For an edge case with -2ν an odd positive integer, 2 square roots of r^ν exist. Potentially, for each solution set, 4 solutions exist.

       Gss.3.2.2   For an edge case with -2ν an odd positive integer, 3 orthogonal   (3.2.44) type-1 solutions exist.

People apply the next items to traditional $D_P=3$, $r>0$ math. [Item (3.1.4)]

$$D_P = 3 \qquad (3.2.45)$$

$$D = 3 \tag{3.2.46}$$
$$S = \nu, \text{ for some non-negative integer } \nu \tag{3.2.47}$$
$$\Omega = \nu(\nu+D-2) = S(S+D_P-2) = S(S+1) \tag{3.2.48}$$
$$2S+1 \text{ angular solutions pertain} \tag{3.2.49}$$

The next items extend traditional $D_P=3$ math. $D \neq D_P$ is allowed, as is $D=D_P$.

$$D_P = 3 \tag{3.2.50}$$
$$-D_P \leq 2\nu, \text{ with } 2\nu \text{ being an integer} \tag{3.2.51}$$
$$\Omega = \nu(\nu+D-2), \text{ for some non-negative integer } D \tag{3.2.52}$$
$$|\Omega| = S(S+D_P-2) = S(S+1), \text{ for some } S \text{ with } 2S \text{ being a non-negative integer} \tag{3.2.53}$$
$$2S+1 \text{ angular solutions pertain} \tag{3.2.54}$$

The next table summarizes results. Regarding the numbers of solutions sets, the leftmost factor of 2 comes from the existence of 2 cases, namely $\eta>0$ and $\eta<0$.

Table 3.2.2   Numbers of sets of solutions and numbers of orthogonal type-1 solutions per set   (3.2.55)

Type	-2ν	Number of solution sets	Orthogonal type-1 solutions per set	
inside	even and >0	2(2S+1)	1	(3.2.56)
inside	odd and >0	2(2S+1)	1	(3.2.57)
edge	even and >0	2(2S+1)	1	(3.2.58)
edge	odd and >0	2(2S+1)	3	(3.2.59)

The next items provide notation for solution sets.

- The symbol `s± denotes a solution set   (3.2.60)
- 2s is an integer, with $-S \leq s \leq S$ (3.2.61)
- ± is + for $\eta>0$ and is − for $\eta<0$ (3.2.62)

The next items provide notation for type-1 solutions.

- The symbol `s±j denotes a type-1 solution   (3.2.63)
- `s± denotes the solution set   (3.2.64)
- j is an integer, with $1 \leq j \leq$ the number of orthogonal type-1 solutions in the   (3.2.65) solution set

$$\sim \sim \sim$$

The next items show a non-traditional solution. This is an inside solution. Here, $S \neq \nu$. We consider this non-traditional solution to correlate with a ground state.

$$D_P = 3 \tag{3.2.66}$$
$$\nu = -1 \tag{3.2.67}$$
$$D = 3 \tag{3.2.68}$$
$$\Omega = \nu(\nu+D-2) = -1(0) = 0 \tag{3.2.69}$$
$$S = 0 \tag{3.2.70}$$
$$\Omega = S(S+1) \tag{3.2.71}$$
$$\xi = (1/2)\, \xi' \tag{3.2.72}$$

$$\Psi(r) \propto r^{-1} \exp(-r^{-2} / (2\eta^2)) \tag{3.2.73}$$

Along with the non-traditional solution, the next items pertain. Traditionally, people state that the S=0 solution below corresponds to the ground state. For the S=0 solution below, $\xi = (3/2)\,\xi'$.

$$\xi = (D + 2\nu)(\xi'/2) = (D_P/2 + \nu)\,\xi' = (3/2 + S)\,\xi' \tag{3.2.74}$$
$$S \text{ is a non-negative integer} \tag{3.2.75}$$
$$\Omega = S(S+1) \tag{3.2.76}$$
$$\Psi(r) \propto r^S \exp(-r^{-2} / (2\eta^2)) \tag{3.2.77}$$

~ ~ ~

Ref.3.2.1   Wolfram Alpha, computational knowledge engine, Wolfram Alpha LLC, http://mathworld.wolfram.com/DeltaFunction.html.

Section 3.3   Numbers of generators for groups SU(j)

Abs.3.3.1   We note relationships between numbers of generators for groups SU(j) for various j.
- This work may show underutilized arithmetic relationships between numbers of generators for various SU(j) groups.

In anticipation of discussing limits on elementary particles within various families, we provide the following information.

The next item shows the number of generators associated with each mathematical group SU(j). Here, j is an integer with j≥2.

- The number of generators for SU(j) is j^2-1 (3.3.1)

The next table shows numbers of generators for some values of j. The term multiple of k refers to multiple of k number of generators. We list integer multiples.

Table 3.3.1   Ratios of numbers of generators for various SU(j)

j	Number of generators	Multiple of 8	Multiple of 24	Multiple of 48	(3.3.2)
3	8	1	-	-	(3.3.3)
5	24	3	1	-	(3.3.4)
7	48	6	2	1	(3.3.5)
9	80	10	-	-	(3.3.6)
11	120	15	5	-	(3.3.7)
13	168	21	7	-	(3.3.8)
15	224	28	-	-	(3.3.9)
17	288	36	12	6	(3.3.10)

80 is not an integer multiple of 48. The next item pertains.

Gss.3.3.1   The number of generators, 48, for SU(7) correlates with some limits (3.3.11)
on solutions that correlate with elementary particles.

The next table pertains.

Table 3.3.2   SU(5) and SU(3)×SU(2)×U(1)
- SU(5) ⊃ SU(3)×SU(2)×U(1)   (3.3.12)
- For SU(3)×SU(2)×U(1), the number of generators is 48   (3.3.13)
  - 48 is the product (8×3×2) of the numbers of generators for the respective groups

The next table shows relationships between numbers of generators.

Table 3.3.3   Other arithmetic relationships between numbers of generators for various SU(j)
- (1 + 2/1) (generators for SU(3)) = (3/1) 8   = 24 = generators for SU(5)   (3.3.14)
- (1 + 1/1) (generators for SU(5)) = (2/1) 24 = 48 = generators for SU(7)   (3.3.15)
- (1 + 2/3) (generators for SU(7)) = (5/3) 48 = 80 = generators for SU(9)   (3.3.16)
- (1 + 1/2) (generators for SU(9)) = (3/2) 80 = 120 = generators for SU(11)   (3.3.17)
- (1 + 2/5) (generators for SU(11)) = (7/5) 120 = 168 = generators for SU(13)   (3.3.18)
- (1 + 1/3) (generators for SU(13)) = (4/3) 168 = 224 = generators for SU(15)   (3.3.19)
- (1 + 2/7) (generators for SU(15)) = (9/7) 224 = 288 = generators for SU(17)   (3.3.20)

Regarding the factors (1 + l) in items above, each of the next items pertains to a series.

- l = 2/1, 2/3, 2/5, and 2/7 pertain respectively for the first, third, fifth, and   (3.3.21)
  seventh relationships above
- l = 1/1, 1/2, and 1/3 pertain respectively for the second, fourth, and sixth   (3.3.22)
  relationships above

Section 3.4   Solutions that might correlate with elementary particles

Abs.3.4.1   We show solutions that might correlate with elementary particles.
- This work provides insight about the use of 3 spatial coordinates for modelling physics.
- This work provides insight regarding relationships between mathematical models for particles and mathematical models for fields.
- This work provides insight regarding which bosons interact with which fermions.

As far as we know, people interpret all experiments and observations as being consistent with the next items.

- Each elementary particle has a spin/ℏ people can denote by S   (3.4.1)
- For each elementary particle, 2S is a non-negative integer   (3.4.2)
- For each known elementary particle, S = 0, 1/2, or 1   (3.4.3)
- People expect that, if gravitons exist, S=2 for gravitons   (3.4.4)

The next items characterize aspects of the set of S that we anticipate to be relevant.

- For each elementary particle, 2S is a non-negative integer   (3.4.5)
- For some elementary particles, S ≥ 3/2   (3.4.6)

People use the next item to model quantum states of physics particles and systems.

$$\Omega = S(S+1) \qquad (3.4.7)$$

Based on items (3.2.26), (3.4.1), (3.4.2), and (3.4.7), the next item defines a relevant number of QP-like dimensions.

$$D_P = 3 \qquad (3.4.8)$$

The next table leads to Table 1.4.1 and Table 1.4.2. Items in the next table do not set upper limits on S. Regarding item (3.4.10), we think that a solution for which the limit $\eta \to 0$ is required might correlate with a particle but not a field. We think that a solution for which any η such that $0<\eta<\infty$ pertains might correlate with a field. For bosons, we think that any solution that correlates with a field correlates with a corresponding particle. (Each of some solutions can correlate with neither a particle nor a field.)

Table 3.4.1   Solutions that might correlate with non-zero-mass elementary particles

Gss.3.4.1   Some non-traditional solutions having $D_P=3$ correlate with the elementary particles, with some properties of elementary particles, and with fields related to elementary particles. (3.4.9)

Gss.3.4.2   For solutions that correlate with non-zero-mass elementary particles, $\nu=-1$ correlates with elementary bosons and their fields, $\nu=-3/2$ correlates with elementary fermion particles, and $\nu=-1/2$ correlates with fermion fields. (3.4.10)

Gss.3.4.3   For elementary particles, $\Omega=+S(S+1)>0$ correlates with QE-like phenomena, $\Omega=0$ correlates with the Higgs boson, and $\Omega=-S(S+1)<0$ correlates with QP-like phenomena. (3.4.11)

- 2S is (3.4.12)
  - an even integer for boson particles
  - an odd integer for fermion particles
- $0 \le S$ (3.4.13)
- $\Omega = \pm S(S+1)$ (3.4.14)
- $\Omega = \nu(\nu+D-2)$, for some integer D>0 (3.4.15)

Gss.3.4.4   For a solution to correlate with an elementary fermion, the relevant $\nu=-3/2$ (that is, particle) solution must correlate with a $\nu=-1/2$ (that is, field) solution. (3.4.16)

Gss.3.4.5   For a solution to correlate with non-zero-mass elementary particles, D for the particles must be a positive integer. (3.4.17)

Gss.3.4.6   For a solution to correlate with non-zero-mass elementary particles, D for the corresponding field must be a positive integer. (3.4.18)

~ ~ ~

For much of the rest of this section, we explore mathematical solutions that might correlate with elementary particles and/or with fields. The next table pertains.

Table 3.4.2   Notes about solutions shown in and discussion flow for this section

- We show some solutions that we think do not correlate with elementary   (3.4.19)
  particles or fields
  - We show these solutions …
    - To provide context for people to gain possibly deeper understanding of
      the math
    - In anticipation of possible uses (for the math) in applications other
      than elementary-particle physics
- We use item (3.3.11) regarding limits on solutions that correlate with   (3.4.20)
  elementary particles [Table 3.4.3]
- Regarding a possible limit of S≤2 (for elementary particles) that people   (3.4.21)
  might say field theory provides, …
  - We at first do not necessarily assume that this aspect of field theory
    pertains
  - We come somewhat close to deriving a limit of S≤2
  - Then, we adopt the limit S≤2
- We explore topics in the following order (Here, for clarity, we use   (3.4.22)
  elementary-particle vocabulary, even though we discuss mathematical
  solutions {including solutions that do not correlate with elementary
  particles}.)
  - Item (3.4.20)
  - Non-zero-mass bosons - the W-, H-, and O-families
  - Fermions - the L- and Q-families
  - Interpreting aspects of representations for L- and Q-family elementary
    particles
  - $\Omega<0$ zero-mass bosons - the Y-family
  - $\Omega>0$ zero-mass bosons - the G-family
  - Instances of sets of elementary particles, based on #E=2 for gravitons
  - Correlations between the Y-, Q-, and O-families
  - Spans for G-family forces
  - Symmetries and numbers of instances regarding peer sets of elementary
    particles
  - Solutions correlating with possible siblings (within a peer) of charged 1L
    elementary particles
  - Number of instances of G-family jjGj& particles
  - Limits on S
  - Number of instances of the H-family particle
  - Number of instances of fermion fields
  - Number of instances of elementary particles
    - This subsection summarizes some work in this section

~ ~ ~

The next table pertains. We interpret item (3.3.11). [Item (3.4.20)] Except to the extent we have criteria that exclude a solution from correlating with an elementary particle, we include solutions for which the number of instances is 48, 24, 8, 4, 2, or 1.

Table 3.4.3   An assumption regarding correlating solutions with elementary particles
        Gss.3.4.7   Regarding solutions that might qualify as correlating with   (3.4.23)
              elementary particles, we exclude solutions for which the number of
              instances is not an even divisor of 48.

~ ~ ~

The next table alludes to solutions that could point to candidates for W-, H-, and O-family elementary particles. For these solutions, $\nu=-1$. For the 2W% solutions and the (2S)0% solutions, $2S \geq \% \geq 0$. We use the term A` to provide a (hexadecimal-like) representation for the (decimal) number 10. The number of particles is $2S+1$. [Item (1.4.6) in Table 1.4.2] Per item (3.4.15), rows with $D \leq 0$ do not correlate with elementary particles.

Table 3.4.4   Solutions possibly correlating with non-zero-mass elementary bosons

S	Ω	D	Traditional particles	Possible particles	Number of particles	(3.4.24)
2	6	−3				(3.4.25)
1	2	1	Z, W⁻, W⁺ (2W%)		3	(3.4.26)
0	0	3	Higgs (0H0)		1	(3.4.27)
1	−2	5		20%	3	(3.4.28)
2	−6	9		40%	5	(3.4.29)
3	−12	15		60%	7	(3.4.30)
4	−20	23		80%	9	(3.4.31)
5	−30	33		A`0%	11	(3.4.32)
...	−S(S+1)					(3.4.33)

~ ~ ~

For W-family solutions, item (1.4.4) limits to S=1 the solutions that are candidates to correlate with elementary particles.

~ ~ ~

For H-family solutions, item (1.4.4) limits to S=0 the solutions that are candidates to correlate with elementary particles.

~ ~ ~

For O-family solutions, possibly relevant (2S)0% solutions exist for arbitrarily large integers S.
The next table shows ground states for candidates for O-family elementary particles with S=2. The number of instances would be 24, which is the number of generators for SU(5). SU(5) correlates with the 5 relevant QP-like oscillators.

Table 3.4.5   O-family solutions that might correlate with S=2 ground states

E 6R	E 6L	E 4R	E 4L	E 2R	E 2L	E 0	P 0	P 2L	P 2R	P 4L	P 4R	P 6L	P 6R	P 8L	P 8R	Solution	(3.4.34) (3.4.35)
	0	#	#	#	#	@	@	@	@	@						404	(3.4.36)
#	0	#	#	#	@	@	@	@	@							403	(3.4.37)
#	#	0	#	#	@	@	@	@	@							402	(3.4.38)
#	#	#	0	#	@	@	@	@	@							401	(3.4.39)
#	#	#	#	0	@	@	@	@	@							400	(3.4.40)

The next table shows ground states for candidates for O-family elementary particles with S=3. The number of instances would be 48, which is the number of generators for SU(7).

Table 3.4.6   O-family solutions that might correlate with S=3 ground states

E 6R	E 6L	E 4R	E 4L	E 2R	E 2L	E 0	P 0	P 2L	P 2R	P 4L	P 4R	P 6L	P 6R	P 8L	P 8R	Solution	
																	(3.4.41)
																	(3.4.42)
0	#	#	#	#	#	#	@	@	@	@	@	@	@			606	(3.4.43)
#	0	#	#	#	#	#	@	@	@	@	@	@	@			605	(3.4.44)
#	#	0	#	#	#	#	@	@	@	@	@	@	@			604	(3.4.45)
#	#	#	0	#	#	#	@	@	@	@	@	@	@			603	(3.4.46)
#	#	#	#	0	#	#	@	@	@	@	@	@	@			602	(3.4.47)
#	#	#	#	#	0	#	@	@	@	@	@	@	@			601	(3.4.48)
#	#	#	#	#	#	0	@	@	@	@	@	@	@			600	(3.4.49)

~ ~ ~

The next table points to solutions that might correlate with L- or Q-family elementary particles. For these solutions, $\nu=-1/2$ for fields and $\nu=-3/2$ for particles. We use item (1.4.4) to compute values of D. The notation 1L denotes various leptons. All leptons have 2S=1. The notation (2S)Q denotes known (for 2S=1) elementary particles or candidate (for 2S≥3) elementary particles. Per items (3.4.15) and (3.4.17), rows with non-integer D for fermion particles do not correlate with elementary particles. Per item (3.4.18), rows with non-positive D for fermion fields do not correlate with elementary particles. We use item (1.4.5) to note that each particle has 3 generations. We use item (1.4.6) to compute, assuming all particles are Dirac particles, numbers of particles per generation and total number of particles. We use item (1.4.7) to compute AUL#MF, an upper limit on the number of Majorana fermions, across 3 generations. For S≥11/2, use of item (1.4.7) would produce a negative number for AUL#MF.

Table 3.4.7   Solutions possibly correlating with non-zero-mass elementary fermions

S	Ω	D for fermion fields	D for fermion particles	Known types of particles	Possible other particles	Number of particles (if all Dirac) • per gen • total	AUL#MF	
								(3.4.50)
3/2	15/4	−5	1					(3.4.51)
1/2	3/4	1	3	Leptons (1L)		• 4 • 12	3	(3.4.52)
1/2	−3/4	4	4	Quarks (1Q)		• 4 • 12	0	(3.4.53)
3/2	−15/4	10	6		3Q	• 8 • 24	6	(3.4.54)
5/2	−35/4	20	28/3					(3.4.55)
7/2	−63/4	34	14		7Q	• 16 • 48	6	(3.4.56)
9/2	−99/4	52	20		9Q	• 20 • 60	0	(3.4.57)
11/2	−143/4	74	82/3					(3.4.58)
13/2	−195/4	100	36			• 28 • 84	?	(3.4.59)

S	Ω	D for fermion fields	D for fermion particles	Known types of particles	Possible other particles	Number of particles (if all Dirac) • per gen • total	AUL#MF	(3.4.50)
...	−S(S+1)							(3.4.60)

~ ~ ~

For L-family solutions, item (1.4.4) limits to S=1/2 the solutions that are candidates to correlate with elementary particles.

The next table pertains. [Table 1.6.7 and Table 1.6.8]

Table 3.4.8   Instances of leptons

Gss.3.4.8   For L-family representations, the number of −1 values correlating (3.4.61) with QE-like oscillators plays a role, similar to the roles played by numbers of @ values for bosons, regarding determining numbers of instances.

Gss.3.4.9   People can overlook a possibility that the number of instances for 1L (3.4.62) Dirac fermions is 48.

- Mathematically, $SU(5) \supset SU(3) \times SU(2) \times U(1)$
- 48 is the number (8×3×2) of generators of $SU(3) \times SU(2) \times U(1)$
- For the L-family, ...   (3.4.63)
  - The number of instances for Dirac fermions is 24, which is the number of generators for SU(5)
  - The number of instances for Majorana neutrinos is 48, which is the number of generators for SU(7)

~ ~ ~

For Q-family solutions, based on work above, possibly (2S)Q% particles exist for arbitrarily large odd positive integers 2S.

We construct representations of how spin-(S"−1/2) fermions (that is, (2S"−1)Q particles) would interact with spin-S" 0-family bosons (that is, (2S")O particles).

An example features quarks (which have S=1/2) and 20% bosons (which would have S"=1). [Regarding this example, see Table 1.6.1, Table 1.6.2, and Table 1.6.4.]

Including and beyond S=1/2, the next discussion shows an algorithm for constructing representations. We note that 4S" equals the number of particles per generation if all the particles are Dirac fermions.

- Construct a table   (3.4.64)
  - Include columns labeled E(2S")R, E(2S")L, ..., E0, P0, ..., P(2S")L, P(2S")R
  - Include 4S" rows
- Fill the table with entries of −1   (3.4.65)
- For the columns P(2S")L and P(2S")R in the table, replace (starting from the (3.4.66) top row) each block of 4 rows with the pattern the 2 P-columns in Table 3.4.9 specify

- For successive blocks of 4 rows (starting from the top row), place the   (3.4.67)
  pattern the 2 E-columns in Table 3.4.9 specify into the table's columns
  E(2S"−2j)R and E(2S"−2j)L
  - Here, j+1 numbers the successive blocks of 4 rows
  - Here, 0≤j<S"

Table 3.4.9   Prototypes for use in representations of interactions involving Q-family fermions

E (2S"−2j)R	E (2S"−2j)L	P (2S")L	P (2S")R	
				(3.4.68)
				(3.4.69)
−2	−1	−1	−2	(3.4.70)
−1	−2	−1	−2	(3.4.71)
−2	−1	−2	−1	(3.4.72)
−1	−2	−2	−1	(3.4.73)

The next table shows results for 3Q. Here, 2S"=4. Each … denotes a −1 that correlates with item (3.4.65). We use this … notation in hopes of making it adequately easy to spot results from steps after item (3.4.65). For each symbol for a particle, the first item in parenthesis correlates with the QE-like oscillator n' for which N(n')=−2. For each symbol for a particle, the second item in parenthesis correlates with the QP-like oscillator n" for which N(n")=−2. The table summarizes interactions with spin-2 O-family bosons (that is, 4O bosons). Each item in the next table satisfies Œ=0. The symbol −1 indicates that no property-changing interaction can occur. The symbol −2 indicates that an interaction-vertex exists. Per Table 3.4.7, up to 6 of these solutions (across 3 generations of 3Q particles) could correlate with Majorana fermions. We think that Majorana fermions correlate with oscillators E2R and E2L. We think that $2^{-1/2}$ (<3Q(2R,4R)| + <3Q(2L,4R)|) could be a relevant destruction operator. We think that $2^{-1/2}$ (<3Q(2R,4L)| + <3Q(2L,4L)|) could be a relevant destruction operator. In Table 3.4.11, we discuss implications (regarding boson-mediated interactions) of the presence throughout some PlL and PlR columns of no values other than −1.

Table 3.4.10   Solutions for 3Q, assuming Dirac fermions and 3 similar generations

E 6R	E 6L	E 4R	E 4L	E 2R	E 2L	E 0	P 0	P 2L	P 2R	P 4L	P 4R	P 6L	P 6R	P 8L	P 8R	Particle or Mode	
																	(3.4.74)
																	(3.4.75)
		−2	−1	…	…	…	…	…	…	−1	−2					3Q(4R,4R)	(3.4.76)
		−1	−2	…	…	…	…	…	…	−1	−2					3Q(4L,4R)	(3.4.77)
		−2	−1	…	…	…	…	…	…	−2	−1					3Q(4R,4L)	(3.4.78)
		−1	−2	…	…	…	…	…	…	−2	−1					3Q(4L,4L)	(3.4.79)
		…	…	−2	−1	…	…	…	…	−1	−2					3Q(2R,4R)	(3.4.80)
		…	…	−1	−2	…	…	…	…	−1	−2					3Q(2L,4R)	(3.4.81)
		…	…	−2	−1	…	…	…	…	−2	−1					3Q(2R,4L)	(3.4.82)
		…	…	−1	−2	…	…	…	…	−2	−1					3Q(2L,4L)	(3.4.83)

~ ~ ~

We summarize some work above regarding elementary fermions and property-changing interactions. The next table pertains. Here, we do not discuss (2S)Q solutions for which S>3/2. Here, we do not discuss 3Q Majorana solutions. We think item (3.7.67) follows from item (3.4.84).

Table 3.4.11   Active and inactive oscillators, regarding representations for elementary fermions

- For elementary fermions, oscillators E0 and P0 do not correlate with   (3.4.84)
  property changes

- For 1L elementary fermions, no EjR or EjL oscillator correlates with property changes (3.4.85)
- For each Dirac elementary fermion, ... (3.4.86)
  - Exactly 1 of oscillators P(#P)R and P(#P)L correlates with property changes
  - The other 1 of oscillators P(#P)R and P(#P)L does not correlate with property changes
  - For any even integer j with 0<j<#P, each of the oscillators PjR and PjL does not correlate with property changes
- For each 1L Majorana elementary fermion, each of oscillators P2R and P2L correlates with property changes (3.4.87)
- For each Q-family Dirac elementary fermion, ... (3.4.88)
  - Exactly 1 EjL or EjR oscillator correlates with property changes
    - Here j = 2 for 1Q and j = 4 or 2 for 3Q
  - The other 1 (for 1Q) or 3 (for 3Q) oscillators do not correlate with property changes

The next items pertain.

Tbd.3.4.1   Describe 3Q properties that correlate with each of oscillator pairs E4R and E4L, E2R and E2L, and P4L and P4R. (3.4.89)
- Possibly, the properties for E4R and E4L and for P4L and P4R correlate with mass (3.4.90)

~ ~ ~

We turn our attention toward the Y-family.

In traditional physics, an early step toward discussion of gluons consisted of people's thinking that a new property (color charge) for quarks was needed so as to correlate with (in nucleons) there being 3 fermions (quarks) in what people considered to be no more than 2 states.

Above, we point to possibilities for 3Q particles. There, the relevant range of oscillators is E4R through P4R. The next table shows some (but not all) solutions that would correlate with ground states for spin-2 Y-family particles. Other solutions would have non-zero values for N(E4R) and/or N(E4L).

Table 3.4.12 Y-family solutions that would correlate with some S=2 ground states

E 6R	E 6L	E 4R	E 4L	E 2R	E 2L	E 0	P 0	P 2L	P 2R	P 4L	P 4R	P 6L	P 6R	P 8L	P 8R	Particle or Mode	
		0	0	-1	0	1	@	@	@	@	@					4Y20	(3.4.93)
		0	0	1	-1	0	@	@	@	@	@					4Y12	(3.4.94)
		0	0	0	1	-1	@	@	@	@	@					4Y01	(3.4.95)
		0	0	0	-1	1	@	@	@	@	@					4Y10	(3.4.96)
		0	0	1	0	-1	@	@	@	@	@					4Y02	(3.4.97)
		0	0	-1	1	0	@	@	@	@	@					4Y21	(3.4.98)

(3.4.91) and (3.4.92) label the two header rows.

Overall, there would be 5 color charges for 3Q fermions. Each such color charge would correlate with 1 of the oscillators E4R through E0.

We consider that the value −1 can pertain to 1 of 5 N(Ej) and, then, that the value +1 can pertain to 1 of 4 remaining N(Ek). Here, 20 = 5×4. The Y-family has 20 4Y (or spin-2) solutions.

For the purposes of this paragraph, we discuss 3Q fermions as if Dirac-fermion pertains. Of the 20 4Y particles, 10 would interact with 3Q fermions for which N(P4L)=−1 and N(P4R)=−2. The other 10 would interact with 3Q fermions for which N(P4L)=−2 and N(P4R)=−1.

The number of relevant QE-like oscillators is 5. The number of generators for SU(5) is 24. Each of the 2 sets of spin-2 gluon-analogs would include 24 particles.

For the S=2 part of the Y-family, people might say that a GRX×U(1) QE-like symmetry applies. Here, GRX denotes a group having 10 generators. The 4Y analog of SU(3)×SU(2)×U(1) is SU(3)×GRX×U(1).

> Tbd.3.4.2   Determine the group GRX for which the QE-like symmetry   (3.4.99)
> GRX×U(1) pertains for 4Y particles.

Paralleling discussion above regarding 1L Dirac fermions, the number of instances of 4Y particles is 24.

~ ~ ~

For the G-family, the next item suggests some math solutions that do not correlate with nature. For these solutions, S=0.

> • Solutions (such as 60G246& and 80G2468&) that would correspond to   (3.4.100)
>   spin-0 G-family particles do not correlate with nature.

The next table shows a possibility for multiple non-zero spins. For 6kG26&, these solutions correlate respectively k=4 and k=8. Item (3.4.103) correlates with S=2. Item (3.4.104) correlates with S=4.

Table 3.4.13 Math solutions correlating with S=2 and S=4 for 6kG26& ground states

E	E	E	E	E	E	E	P	P	P	P	P	P	P	P	P	Particle or	(3.4.101)
6R	6L	4R	4L	2R	2L	0	0	2L	2R	4L	4R	6L	6R	8L	8R	Mode	(3.4.102)
		@	@	@	-2	0	#	#	#	#	0					64G2L6R	(3.4.103)
		@	@	@	-2	0	#	#	#	0	#					68G2L6L	(3.4.104)

The next item pertains.

> Gss.3.4.10   For any values of j and %, to the extent multiple values of k could   (3.4.105)
> pertain, any jkG%& solution that correlates with an elementary particle
> correlates with the minimal positive value of k. Here, k=2S.

The next table shows candidates for ground states for G-family bosons for which 2≤#P≤8. Each of these candidates satisfies the criterion Table 3.4.3 provides. For the items for which columns E2R and E2L show blanks (that is, the #E=0 G-family solutions), we discuss numbers of instances below. [Table 3.4.18 and item (3.4.205)] Of the candidates this table shows, only 66G6&, 86G6&, and 88G8& would have S>2.

Table 3.4.14 Ground states for #P≤8 G-family solutions that might correlate with elementary particles

E	E	E	E	E	E	E	P	P	P	P	P	P	P	P	P	Solution	(3.4.106)
6R	6L	4R	4L	2R	2L	0	0	2L	2R	4L	4R	6L	6R	8L	8R		(3.4.107)
					@	-1	0	0								22G2&	(3.4.108)
		@	@	@	-1	0	0	#	#							42G2&	(3.4.109)
		@	@	@	-1	#	#	0	0							44G4&	(3.4.110)
				@	-2	0	0	0	0							42G24&	(3.4.111)
@	@	@	@	@	-1	0	0	#	#	#	#					62G2&	(3.4.112)
@	@	@	@	@	-1	#	#	0	0	#	#					64G4&	(3.4.113)

E6R	E6L	E4R	E4L	E2R	E2L	E0	P0	P2L	P2R	P4L	P4R	P6L	P6R	P8L	P8R	Solution	
																	(3.4.106)
																	(3.4.107)
		@	@	@	@	@	−1	#	#	#	#	0	0			66G6&	(3.4.114)
			@	@	@	@	−2	0	0	0	0	#	#			62G24&	(3.4.115)
			@	@	@	@	−2	0	0	#	#	0	0			64G26&	(3.4.116)
			@	@	@	@	−2	#	#	0	0	0	0			62G46&	(3.4.117)
						@	−3	0	0	0	0	0	0			64G246&	(3.4.118)
@	@	@	@	@	@	@	−1	0	0	#	#	#	#	#	#	82G2&	(3.4.119)
@	@	@	@	@	@	@	−1	#	#	0	0	#	#	#	#	84G4&	(3.4.120)
@	@	@	@	@	@	@	−1	#	#	#	#	0	0	#	#	86G6&	(3.4.121)
@	@	@	@	@	@	@	−1	#	#	#	#	#	#	0	0	88G8&	(3.4.122)
		@	@	@	@	@	−2	0	0	0	0	#	#	#	#	82G24&	(3.4.123)
		@	@	@	@	@	−2	0	0	#	#	0	0	#	#	84G26&	(3.4.124)
		@	@	@	@	@	−2	0	0	#	#	#	#	0	0	86G28&	(3.4.125)
		@	@	@	@	@	−2	#	#	0	0	0	0	#	#	82G46&	(3.4.126)
		@	@	@	@	@	−2	#	#	0	0	#	#	0	0	84G48&	(3.4.127)
			@	@	@	@	−2	#	#	#	#	0	0	0	0	82G68&	(3.4.128)
				@	@	@	−3	0	0	0	0	0	0	#	#	84G246&	(3.4.129)
				@	@	@	−3	0	0	0	0	#	#	0	0	82G248&	(3.4.130)
				@	@	@	−3	0	0	#	#	0	0	0	0	84G268&	(3.4.131)
				@	@	@	−3	#	#	0	0	0	0	0	0	82G468&	(3.4.132)
						@	−4	0	0	0	0	0	0	0	0	84G2468&	(3.4.133)

~ ~ ~

Above, we show that #E=2 for gravitons. An SU(3) symmetry pertains regarding QE-like matters. SU(3) has 8 generators. The next table pertains.

Table 3.4.15   8 gravitons and 8 peers of stuff

Gss.3.4.11 The set of elementary particles includes 8 instances of gravitons.		(3.4.134)
Each instance of gravitons correlates with a set of elementary fermions.		
Each such set of elementary fermions is identical to each other set.		
• We use the term peer for each set		(3.4.135)
• We the denote the peers by MP(n), with 0≤n≤7		
• We number the peers so that MP(0) includes baryonic matter		

~ ~ ~

We correlate Y-family members having S=S" with Q-family members having S=S"−(1/2). For the O-family, Ω<0. O-family bosons could only be observed (by QE-like means) in groups, much as Q-family fermions can only be observed (by QE-like means) in groups (such as in the forms of pairs in mesons and such as in the forms of triples in baryons). The next table pertains.

Table 3.4.16   Correlations among Y-, Q-, and O-family elementary particles

Gss.3.4.12 O-family bosons with spin-S are bound into groups by Y-family members with spin-S.		(3.4.136)
Gss.3.4.13 The minimum number of O-family bosons in a bound group containing a (2S)O boson is 2S+1.		(3.4.137)

Gss.3.4.14   For S" a positive integer and S=S"−(1/2), either all of the (2S")O,   (3.4.138)
    (2S")Y, and (2S)Q solutions correlate with elementary particles or none of
    those solutions correlates with an elementary particle.

Gss.3.4.15   The number of instances of (2S)Q equals the number of instances of   (3.4.139)
    (2S")Y, for S=S"−(1/2).

Gss.3.4.16   The number of instances of (2S)Q equals the number of instances of   (3.4.140)
    (2S")O, for S=S"−(1/2).

~ ~ ~

We return to discussion of G-family forces and of instances. The next table pertains. [Table 3.4.15]

Table 3.4.17   Spans between peers, for instances of G-family forces

• For the next items, 0≤n≤7, 0≤n'≤7, and n≠n'	(3.4.141)

Gss.3.4.17   No 2kG%& force mediates an interaction between a fermion in   (3.4.142)
    MP(n) and a fermion in MP(n').

Gss.3.4.18   No 4kG%& force mediates an interaction between a fermion in   (3.4.143)
    MP(n) and a fermion in MP(n').

Gss.3.4.19   6kG%& forces can mediate interactions between fermions in MP(n)   (3.4.144)
    and fermions in MP(n'), for appropriate choices of n and n'.

Gss.3.4.20   8kG%& forces can mediate interactions between fermions in MP(n)   (3.4.145)
    and fermions in MP(n'), for appropriate choices of n and n'.

Gss.3.4.21   Items (3.4.147), (3.4.148), and (3.4.149) define, in effect, the term   (3.4.146)
    appropriate in items (3.4.144) and (3.4.145).

  • We number the peers so that …   (3.4.147)
    • For each of n = 0, 2, 4, or 6, an instance of 64G246& mediates interactions
      between fermions in MP(n) and fermions in MP(n+1)
    • For each of n = 0 or 4, an instance of 84G2468& mediates interactions
      between fermions in MP(n) and fermions in MP(n+1), MP(n+2), and
      MP(n+3)

  • For n = 0, 2, 4, or 6 and n' = 0, 2, 4, or 6, an instance of 64G246& that   (3.4.148)
    mediates interactions involving fermions in MP(n) and MP(n+1) does not
    mediate interactions involving fermions in MP(n') or MP(n'+1)

  • For n = 0 or 4 and n' = 0 or 4, an instance of 84G2468& that mediates   (3.4.149)
    interactions involving fermions in MP(n) through MP(n+3) does not
    mediate interactions involving fermions in MP(n') through MP(n'+3)

  • Each jkG%& force can mediate interactions between fermions in MP(n)   (3.4.150)

The next table pertains.

Table 3.4.18   Number of instances for some #E=0 G-family members

Solution	Number of instances	
		(3.4.151)
42G24&	8	(3.4.152)
64G246&	4	(3.4.153)
84G2468&	2	(3.4.154)

~ ~ ~

In this subsection, we explore limits on numbers of instances.

The next table notes symmetries that correlate with work in Part 1 and Part 3. We use Table 3.4.3. Each row pertains to a class of particles. For each row, the symmetry type correlates with the 1 of QE-like and QP-like that correlates with numbers of instances of the types of particles (and does not correlate with details of interaction vertices in which the particles participate). The instances of particles column shows the number of generators for the counting-related symmetry. In the instances per peer column, we, in effect, allocate the instances equally across MP(n) peers. Per Table 3.4.16, we assign the same number of instances to 1Q as we assign to 2Y. Per Table 3.4.16, we assign the same number of instances to 3Q as we assign to 4Y.

Table 3.4.19 Symmetries related to determining numbers of instances for some elementary particles

Symmetry type	Counting-related symmetry	Type of particle	Instances of particles	Instances per peer	(3.4.155)
QE-like	SU(3)	44G4& (graviton)	8	1	(3.4.156)
QP-like	SU(3)	2Y (Y-family)	8	1	(3.4.157)
QP-like	SU(3)	2O (O-family)	8	1	(3.4.158)
from 2Y	SU(3)	1Q (Q-family)	8	1	(3.4.159)
QE-like	SU(3)	2W (W-family)	8	1	(3.4.160)
QE-like	SU(5)	1L (L-family) Dirac fermions	24	3	(3.4.161)
QE-like	SU(7)	1L (L-family) Majorana neutrinos	48	6	(3.4.162)
QP-like	SU(5)	4Y (Q-family)	24	3	(3.4.163)
QP-like	SU(5)	4O (O-family)	24	3	(3.4.164)
from 4Y	SU(5)	3Q (Q-family)	24	3	(3.4.165)

~ ~ ~

In Table 3.4.17, we use QE-like symmetry to point to possible peers we label as MP(n). In Table 3.4.19, we point to a possibility that 3 or 6 sets of 1L particles pertain for each peer. In this subsection, we use QP-like symmetry to point to possible siblings of baryonic matter 1L elementary particles.

The next table pertains. Here, we think that, for BMS6(1), BMS6(3), and BMS6(5) charged-lepton solutions, the property-changing absorption pattern would correlate with the only absorption by a positively (or, respectively, negatively) charged lepton being that of Q=+1 (Q=−1). We know of no elementary fermions for which |Q'|=2. Alternatively, people might say that right-handed matter (and left-handed antimatter) charged leptons do not exist. [Table 3.4.22]

Table 3.4.20 Notation for siblings of baryonic-matter charged 1L elementary particles

- BMS6(j) denotes possibilities for solutions corresponding to siblings of baryonic-matter charged 1L elementary particles (3.4.166)
  - Here, $0 \leq j \leq 5$
  - Here, BMS6(0) correlates with baryonic matter
- For charged leptons, BMS(j=even) sibling solutions do not correlate with elementary particles (3.4.167)
- BMS3(j) correlates with pairs of BMS6(k) sibling solutions (3.4.168)
  - Here, $0 \leq j \leq 2$
  - BMS3(0) correlates with BMS6(0)+BMS6(1)
  - BMS3(1) correlates with BMS6(2)+BMS6(3)
  - BMS3(2) correlates with BMS6(4)+BMS6(5)

The next table shows an orderly array of approximate particle masses for quarks and charged leptons. For each particle, an item shows $\log_{10}(\text{mass}/m_e)$, charge in units of Q', and particle name. (In Section 4.4, we discuss an approximate formula for masses for known 1Q and charged 1L elementary particles. In terminology just introduced, these charged 1L elementary particles are the BMS6(0) elementary particles.)

Table 3.4.21 A mass-centric periodic-table-like array for quarks and BMS6(0) elementary particles

M'' \ M'	−3	−2	−1	0	1	...	
							(3.4.170)
0	0.00 (−1) electron	0.61 (+2/3) up	0.97 (−1/3) down		0.97 (+1/3) anti-down		(3.4.171)
1		2.26 (−1/3) strange	3.40 (+2/3) charm		3.40 (−2/3) anti-charm		(3.4.172)
2	2.32 (−1) muon	3.93 (−1/3) bottom	5.51 (+2/3) top		5.51 (−2/3) anti-top		(3.4.173)
3	3.54 (−1) tauon						(3.4.174)

The next table shows how to extend the above table from 1 sibling to 6 siblings. Here, we show 6 groups of solutions. We label the 6 sibling groups via the j for BMS6(j). For 1Q, each candidate sibling resides within 3 contiguous rows, with 1 generation per row. (Per item (3.4.159), people might say that the BMS6(j>0) items in the 1Q column correlate with a pattern but not with elementary particles.) For 1L, each sibling resides within 4 contiguous rows, with 1 generation for each of 3 rows and no solutions for the other 1 row. The concept columns extend cyclic labelling, based on names of leptons for which 0≤M''≤3. Here, positron denotes a standard term for a-electron. Here, a- denotes anti. People might say that each group (of the 6 groups of solutions) corresponds to 1 of the 6 instances that correlates with the SU(2)×U(1) symmetry for 1L solutions. We label rows by n` for instances of U(1). There are 2 such instances. We label rows by k for show instances of SU(2). There are 3 such instances. For k for SU(2):k entries of the form 1, 2 (or 2, 3), the first number in a pair pertains to the leftmost concept column (n`=1) and the second number in a pair pertains the rightmost concept column (n`=2).

Table 3.4.22 Sets of quark solutions and charged-lepton solutions for the peer MP(0)

M''	1Q j, g j for BMS6(j) g for gen	charged 1L j for BMS6(j)	charged 1L concept (n`=1)	charged 1L concept (n`=2)	charged 1L n` for U(1):n`	charged 1L k for SU(2):k	
							(3.4.176)
0	0, 1	0	electron		1	1	(3.4.177)
1	0, 2					1	(3.4.178)
2	0, 3	0	muon		1	1	(3.4.179)
3	(1, 1)	0, 1	tauon	a-tauon	1, 2	1	(3.4.180)
4	(1, 2)					1	(3.4.181)
5	(1, 3)	1		positron	2	1	(3.4.182)
6	(2, 1)	1, 2	muon	a-muon	1, 2	1, 2	(3.4.183)
7	(2, 2)					2	(3.4.184)
8	(2, 3)	2	tauon		1	2	(3.4.185)

M"	1Q	charged 1L	charged 1L	charged 1L	charged 1L	charged 1L	(3.4.175)
	j, g j for BMS6(j) g for gen	j for BMS6(j)	concept (n`=1)	concept (n`=2)	n` for U(1):n`	k for SU(2):k	(3.4.176)
9	(3, 1)	2, 3	electron	positron	1, 2	2	(3.4.186)
10	(3, 2)					2	(3.4.187)
11	(3, 3)	3		a-muon	2	2	(3.4.188)
12	(4, 1)	3, 4	tauon	a-tauon	1, 2	2, 3	(3.4.189)
13	(4, 2)					3	(3.4.190)
14	(4, 3)	4	electron		1	3	(3.4.191)
15	(5, 1)	4, 5	muon	a-muon	1, 2	3	(3.4.192)
16	(5, 2)					3	(3.4.193)
17	(5, 3)	5		a-tauon	2	3	(3.4.194)
18	-	5		positron	2	3	(3.4.195)

The above table does not include neutrinos.

In Section 4.4, we discuss charges and masses for the possible 5 BMS6(j>0) charged-lepton sibling solutions. In Section 4.4, we discuss neutrinos.

~ ~ ~

In this subsection, we explore possibilities regarding numbers of instances for jjGj& G-family solutions. The next table shows how QE-like symmetries might pertain.

Table 3.4.23 Mathematically, possible spans for some G-family solutions

Force: • jjGj& • S=spin	QE-like symmetry	Number of instances	Candidate span of an instance of the solution (whether or not nature exhibits such a force)	(3.4.196)
• 44G4& • S=2	SU(3)	8	• 1 peer • That is, MP(n), for some n, 0≤n≤8	(3.4.197)
• 66G6& • S=3	SU(5)	24	• 2 BMS6 siblings (= 1 BMS3 sibling) within 1 peer	(3.4.198)
• 88G8& • S=4	SU(7)	48	• 1 BMS6 sibling within 1 peer	(3.4.199)
• A`A`GA`& • S=5	SU(9) or other	≥80	• (no such force)	(3.4.200)

The next table lists possibilities for spans for the photon solution. People might say that item (3.4.202) correlates with a lack of an applicable QE-like symmetry. People might say that item (3.4.203) correlates with symmetries pertaining to the W-family and/or to 44G4&. People might say that item (3.4.204) correlates the Standard Model SU(3)×SU(2)×U(1) symmetry.

Table 3.4.24 Alternatives regarding spans for 22G2& solutions

Force: • 22G2& • S=1	Symmetry	Number of instances	Candidate span of an instance of the solution	(3.4.201)

• 22G2&
  • S=1
	None	1	• 8 peers
			  • That is, MP(n) for all n, 0≤n≤8
				(3.4.202)

• 22G2&
  • S=1
	SU(3)	8	• 1 peer
			  • That is, MP(n), for some n, 0≤n≤8
				(3.4.203)

• 22G2& • S=1	SU(3)×SU(2)×U(1)	48	• 1 sibling within 1 peer	(3.4.204)

Based on item (6.3.12), the next item pertains. People might correlate some of the relevance of item (6.3.12) with known physics (as opposed to correlating the item purely with mathematics).

> • The number of instances of 22G2& is 8   (3.4.205)

~ ~ ~

Regarding the topic of the possible existence of G-family particles with S>2, the next table pertains.

Table 3.4.25 Discussion of the possibility of elementary particles with S>2

> • People might say that the following factors favor not correlating G-family   (3.4.206)
>   solutions having S>2 with candidate elementary particles
>   • Traditional field theory
>   • The possibility for many such particles
>   • For example, possibly 46G24& (which would have S=3) would exist (as
>   well as 42G24&, which would have S=1)
>   • Modes for 46G24& would be 46G2L4L and 46G2R4R
>   • The span of an instance of 66G6& or 88G8& would be less than 1 peer
> • For the purposes of this paper, ...   (3.4.207)
>   • We assume that all elementary particles have S≤2
>   • We continue to explore math related to the solutions 66G6&, 86G6&, and
>   88G8&

~ ~ ~

Which of the numbers 1, 8, 24, or 48 correlates with the number of instances of the H-family? People might say that possibly an answer is not required for this paper. We think we find 1 possible effect that correlates with an answer. [Table 5.1.2]

Given concepts above, the next item pertains.

> Gss.3.4.22 The number of instances of 0H0 is 1.   (3.4.208)

~ ~ ~

In this subsection, we discuss fermion fields.

We note items (3.2.57) and (3.2.59). For a given combination of S and Ω, as many field-oriented solution sets exist as do particle-oriented solution sets. The number of field-oriented solutions is 2(2S+1). The next table pertains, based on comparing the number of field-correlated solutions and the number of particle-correlated solutions.

Table 3.4.26 Relationships between fermion particles, fermion fields, and generations
- Fermion fields do not correlate with generation   (3.4.209)
- For Dirac fermions, fermion pair creation (or pair annihilation) that   (3.4.210)
  correlates with fields does not correlate with generation

$$\sim \sim \sim$$

The next table pertains.

Table 3.4.27 Possibilities regarding numbers of instances of elementary particles
- One of 8 or 24 instances of 1L Dirac elementary particles exist   (3.4.211)
- One of 0, 8, or 48 instances of 1L Majorana neutrinos exist   (3.4.212)
- 8 instances of 2Y elementary particles exist   (3.4.213)
- 8 instances of 2O and 1Q elementary particles exist   (3.4.214)
- 8 instances of 2W elementary particles exist   (3.4.215)
- One of 0 or 24 instances of 4Y particles exist   (3.4.216)
- One of 0 or 24 instances of 4O and 3Q elementary particles exist   (3.4.217)
- The number of instances of jkG%& varies by specific particle   (3.4.218)
- 1 instance of the 0H0 elementary particle exists   (3.4.219)

Section 3.5   Some properties and lengths

Abs.3.5.1 The ratio of strengths of electromagnetism and gravity correlates with the ratio of masses of the tauon and the electron.
- This work reduces the number of potentially independent fundamental physics constants.

Abs.3.5.2 We provide candidates for phenomena to which jkG...6...& forces couple and to which jkG...8& forces couple.
- This work points to opportunities determine the next (and also, the last) 2 elements in the series charge, mass, ...

The next item characterizes the relative strengths of electromagnetism and gravity. Here, q_e denotes the charge of an electron, $1/(4\pi\varepsilon_0)$ denotes the Coulomb constant, G_N denotes the gravitational constant, and m_e denotes the mass of an electron. This calculation pertains for electrons and positrons. We base numbers (here and below) on data Section 10.5 shows. The uncertainty-range is approximate.

$$\{(q_e)^2/(4\pi\varepsilon_0)\} / \{G_N(m_e)^2\} \approx 4.1649(\sim1)\times10^{42} \qquad (3.5.1)$$

The next items define β and β'. Here, m_{tauon} denotes the mass of a tauon.

$$(4/3)(\beta^6)^2 = \{(q_e)^2/(4\pi\varepsilon_0)\} / \{G_N(m_e)^2\} \qquad (3.5.2)$$
$$\beta' = m_{tauon} / m_e \qquad (3.5.3)$$

The next items estimate β and β'. For β, we estimate an uncertainty-range based on the uncertainty-range item (3.5.1) shows. (The uncertainty range item (3.5.4) shows may be an over-estimate.) For β', we base the uncertainty-range on experimental results.

$$\beta \approx 3.477139(\sim94)\times10^3 \tag{3.5.4}$$
$$\beta' \approx 3.47715(31)\times10^3 \tag{3.5.5}$$

The next item pertains.

Gss.3.5.1 $\beta' = \beta$. $\tag{3.5.6}$

The next item interprets elements of the left side of item (3.5.2). Here, people might say that the channel that pertains for photons and does not pertain for gravitons correlates with oscillators E2R, E2L, P4L, and P4R. We discuss the concept of channels in Section 8.2.

Gss.3.5.2   In the expression $(4/3)(\beta^6)^2 = \{(q_e)^2/(4\pi\varepsilon_0)\} / \{G_N(m_e)^2\}$, the   (3.5.7)
leftmost exponent 2 represents the number of vertices in a Feynman diagram, β^6 represents the ratio of strengths per channel for electromagnetism and gravity (for interactions between 2 electrons), 4 represents the number of channels for a photon, and 3 represents the number of channels for a graviton.

~ ~ ~

The next item notes a traditional physics length, the Planck length. We provide a symbol, R_2, for that length. Here, ℏ denotes Planck's constant (reduced). Here, c denotes the speed of light. We add to the traditional statement 2 factors, each of value 1. The first such factor is m_e^0. The second such factor is 2^0.

$$R_2 = G_N^{1/2}\, m_e^0\, \hbar^{1/2}\, c^{-3/2}\, 2^0 \tag{3.5.8}$$

The next item applies a traditional formula to a property (mass) associated with electrons. The formula represents the Schwarzschild radius. Traditionally, people apply the Schwarzschild-radius formula to black holes. Traditionally, people do not apply the formula to objects people claim have not enough mass to form black holes. We add to the traditional statement 1 factor with value 1. That factor is $\hbar^0$.

$$R_4 = G_N^1\, m_e^1\, \hbar^0\, c^{-2}\, 2^1 \tag{3.5.9}$$

The next item shows the ratio of the above 2 lengths.

$$Z = R_2 / R_4 = G_N^{-1/2}\, m_e^{-1}\, \hbar^{1/2}\, c^{1/2}\, 2^{-1} \approx 1.1945\times10^{22} \tag{3.5.10}$$

The next item defines a series of lengths.

$$R_j = R_2\cdot Z^{(2-j)/2}, \text{ with j being an even integer} \tag{3.5.11}$$

The next table show factors and values (for electrons and positrons). These approximate lengths are the products of the factors indicated by the five columns having labels k for l^k (for some l). Times are computed via time=length/c. The time-centric column shows the log-base-10 of times (in seconds). The time since the big bang is $\sim10^{17.6}$ seconds. The j column values indicate possibly interesting correlations between items and G-family solutions we symbolize by jjGj&. (We do not define a 00G0& solution.) The j' column values indicate possibly interesting correlations between items and properties of objects. In effect, j'=j+2.

Table 3.5.1   A series of lengths

j	R_j	Length (m)	$\log_{10}$ (time (sec))	Concept	k for $G_N{}^k$	k for $m_e{}^k$	k for $\hbar^k$	k for c^k	k for 2^k	j'	(3.5.12)
		3.3×10^{53}	+45		−1.5	−4	2.5	0.5	−4		(3.5.13)
		2.7×10^{31}	+23		−1	−3	2	0	−3		(3.5.14)
		2.3×10^{9}	+0.88		−0.5	−2	1.5	−0.5	−2		(3.5.15)
0	R_0	1.9×10^{-13}	−21		0	−1	1	−1	−1	2	(3.5.16)
2	R_2	1.6×10^{-35}	−43	Planck length	0.5	0	0.5	−1.5	0	4	(3.5.17)
4	R_4	1.4×10^{-57}	−65	Schwarzschild radius	1	1	0	−2	1	6	(3.5.18)
6	R_6	1.1×10^{-79}	−87		1.5	2	−0.5	−2.5	2	8	(3.5.19)
8	R_8	9.5×10^{-102}	−109.5		2	3	−1	−3	3		(3.5.20)
		7.9×10^{-124}	−132		2.5	4	−1.5	−3.5	4		(3.5.21)

The next item pertains.

> Gss.3.5.3   We attach significance to R_j for which a particle property has an exponent k=0. (3.5.22)

G_N is not a particle property. We note item (3.5.2).

> Gss.3.5.4   Regarding R_0, people can consider q_e to be a particle property for which $|q_e|^0$ pertains. (3.5.23)

The next table pertains. The leftmost 3 columns pertain to electrons. Here, g_S denotes magnetic moment. For electrons, $g_S=2$. The rightmost 2 columns provide names for physical properties. For the column labelled our term ..., PR refers to property, G refers to the G-family, and the number refers both to j' (in Table 3.5.1) and to an oscillator pair (2 for P2L and P2R, 4 for P4L and P4R, 6 for P6L and P6R, and 8 for P8L and P8R). We note that, for PR66G6 and PR88G8, the math solutions pertain, even if the forces do not pertain.

Table 3.5.2   The series charge, mass, ...

j'	Factor (for an electron) for which k=0	Size of the property (for an electron)	Traditional name of the property	Our term for a name for the property	(3.5.24)
2	q_e	q_e	Charge	PR22G2	(3.5.25)
4	m_e	m_e	Mass	PR44G4	(3.5.26)
6	$\hbar$	$(g_S)\hbar/2$ or $2S\hbar$	-	PR66G6	(3.5.27)
8	-	-	-	PR88G8	(3.5.28)

We think that expressions above for the R_j and for Z might contain another factor. We multiply the expression above for R_8 by $(PR88G8)^k$, with k=0. So as not to change the values of the other R_j, we assume that for electrons and positrons PR88G8=1.

The next table shows candidates for PR66G6 of an object.

Table 3.5.3   Candidates for PR66G6 of an object

• Candidates	(3.5.29)
• Magnetic moment for the object	(3.5.30)
• 2S$\hbar$ for the object	(3.5.31)
• Other	(3.5.32)

The next table shows candidates for PR88G8 of an object. Possibly, PR88G8 ≠ PR66G6/$\hbar$.

Table 3.5.4   Candidates for PR88G8 of an object

• Candidates	(3.5.33)
• The object's (magnetic moment)/$\hbar$	(3.5.34)
• 2S for the object	(3.5.35)
• The sum over fermions composing the object of 2S for each fermion	(3.5.36)
• The object's total number of fermions	(3.5.37)
• Other	(3.5.38)

~ ~ ~

Possibly the formula for R_0 pertains to other than electrons and positrons. For Z and W bosons, R_0 may have significance. For pions, R_0 may have significance. The next table pertains.

Table 3.5.5   Concepts indicating possible significance for lengths R_0

- • A charged pion R_0 would be ~0.70×10^{-15} meters   (3.5.39)
  - • That length is a factor ~139.6/0.511 or ~273.2 smaller than that for electrons
  - • An experimental charge radius for charged pions is $0.78 \, ^{+0.09} \, _{-0.10} \times 10^{-15}$ meters [Ref.3.5.1]
- • A Z-boson R_0 is ~2×10^{-18} meters   (3.5.40)
  - • People measure spatial dependence for interactions mediated by the weak interaction
  - • For a separation of ~10^{-18} meters between 2 interacting particles, the weak interaction and the electromagnetic interaction have similar magnitudes [Ref.3.5.2]
  - • At a separation of ~3×10^{-17} meters, the weak interaction is less by approximately a factor of 10^4 [Ref.3.5.2]
  - • A W-boson R_0 is ~$(1.1) \times (R_0$ for a Z boson)

~ ~ ~

We explore the extent to which PR66G6 and PR88G8 might be negative for elementary particles or other objects. The next items pertain. [Table 3.4.23]

Gss.3.5.5   The U(1) in the W-family symmetry SU(3)×SU(2)×U(1) correlates   (3.5.41)
with PR22G2 being (across elementary particles) positive for some
elementary particles and being negative for some elementary particles.

Gss.3.5.6   PR66G6 can be positive or zero (but not negative), depending on the (3.5.42)
elementary particle or object being characterized.

- • Here, the relevant symmetry is SU(5)

Gss.3.5.7   PR88G8 can be positive or possibly zero (but not negative),   (3.5.43)
depending on the elementary particle or object being characterized.
- Here, the relevant symmetry is SU(7)

~ ~ ~

Ref.3.5.1   G. T. Adylov, et. al., A measurement of the electromagnetic size of the pion from direct elastic pion scattering data at 50 GeV/c, *Nuclear Physics B*, Volume 128, Issue 3, 3 October 1977, pages 461-505. (http://dx.doi.org/10.1016/0550-3213(77)90056-6)
Ref.3.5.2   Particle Data Group, Electroweak (web page), *The Particle Adventure*, Lawrence Berkeley National Laboratory, http://www.particleadventure.org/electroweak.html.

Section 3.6   Solutions correlating with possible elementary particles

Abs.3.6.1   We summarize possible elementary particles that correlate with solutions.
- This work provides themes for experimental research.

The next table lists spins for elementary particles. This table summarizes results from Section 3.4. In the spin* column, we allude to elementary particles that people might say lie beyond the Standard Model. (We predict various currently not known G-family elementary particles, such as 42G24&, that would have spin 1.) The table notes the possibility of siblings for leptons. The table notes that, except for Higgs bosons, all particles have peers. The table does not note possible associations (within a peer) of sets of 4O, 3Q, and 4Y particles with lepton sibling groups.

Table 3.6.1   Spins for possible elementary particles

Family	Spin	Spin*	Siblings	Peers	
G-	1	1 , 2		Yes	(3.6.2)
W-	1			Yes	(3.6.3)
H-	0				(3.6.4)
O-	-	1 , 2		Yes	(3.6.5)
L-	1/2		Possible	Yes	(3.6.6)
Q-	1/2	3/2		Yes	(3.6.7)
Y-	1	2		Yes	(3.6.8)

(3.6.1)

This next items point to possibilities for experimental or observational physics. The term interacts directly with denotes has an interaction vertex with. Item (3.6.12) correlates with Table 3.4.22. Item (3.6.13) correlates with Table 3.4.23.

Tbd.3.6.1   To what extent does nature exhibit a set of particles (for each peer)   (3.6.9)
correlating with the collection 4O, 3Q, and 4Y?
Tbd.3.6.2   Assuming 4O, 3Q, and 4Y particles exist, to what extent do 4O   (3.6.10)
bosons interact directly with 1Q fermions?
Tbd.3.6.3   Assuming 4O, 3Q, and 4Y particles exist, to what extent do 2O and   (3.6.11)
2W bosons interact directly with 3Q fermions?
Tbd.3.6.4   Assuming that (for 1 peer) more than 1 sibling exists and assuming   (3.6.12)
that multiple instances of 4O, 3Q, and 4Y pertain (for that 1 peer), to what
extent do siblings and instances of 4O, 3Q, and 4Y correlate?

Tbd.3.6.5   To what extent can observations or experiments detect, infer, or   (3.6.13)
rule out (to some confidence level) S>2 G-family bosons (such as might
correlate with solutions 66G6&, 86G6&, and 88G8&)?

Possibly, answers to items (3.6.10) and (3.6.11) are to no extent.

Section 3.7   Mathematics related to kinematics

Abs.3.7.1   We discuss quantum operators related to motion of elementary particles.
- This work provides a basis for extending our interaction-centric approach (to modeling
  particle physics) toward a traditional approach to modeling particle kinematics.

Abs.3.7.2   We find quantum numbers related to mass.
- This work provides new insight regarding masses of elementary particles.

Abs.3.7.3   For non-zero-mass elementary particles, $E^2 - c^2P^2 = \text{sign}'(\Omega)\, |m^2|\, c^4$.
- This work correlates with the possibility that the quantum mechanics of quarks or spin-3/2
  fermions correlates with inflation.

Above, we do not much address details of quantum kinematics of elementary particles or other objects.
For $D_P=3$, the next items provide an alternative form of the Laplacian operator. Here, r_2 provides a radial
coordinate in 2 dimensions, φ provides the related angular coordinate, and x provides a linear coordinate
for the third dimension. We use coordinates y and z for a linear description correlating with $r_2{}^2$. [Ref.3.7.1]

$$\nabla^2 = r_2^{-1}(\partial/\partial r_2)(r_2)(\partial/\partial r_2) + r_2^{-2}(\partial^2/\partial^2\varphi) + (\partial^2/\partial^2 x) \quad (3.7.1)$$
$$r_2{}^2 = y^2 + z^2 \quad (3.7.2)$$

The next items restate item (3.7.1).

$$\nabla^2 = \nabla_2{}^2 + (\partial^2/\partial^2 x) \quad (3.7.3)$$
$$\nabla_2{}^2 = r_2^{-1}(\partial/\partial r_2)(r_2)(\partial/\partial r_2) + r_2^{-2}(\partial^2/\partial^2\varphi) \quad (3.7.4)$$

The next items provide relevant equations and solutions for $D_{P'}=2$.

$$\xi\,\Psi(r_2) = (\xi'/2)\,(\,-\eta^2\,\nabla_2{}^2 + \eta^{-2}r_2{}^2\,)\,\Psi(r_2) \quad (3.7.5)$$
$$\nabla_2{}^2 = r_2^{-1}(\partial/\partial r_2)(r_2)(\partial/\partial r_2) - \Omega_2 r_2^{-2} \quad (3.7.6)$$
- ξ and $\xi'/2$ denote numbers   (3.7.7)
- $\Psi(r_2)$ denotes a wave function   (3.7.8)
- r_2 denotes a variable, with dimensions of length   (3.7.9)
- η denotes a length   (3.7.10)
- Ω_2 denotes a number   (3.7.11)

$$V = (\xi'/2)\,\eta^{-2}\,r_2{}^2 \quad (3.7.12)$$

Paralleling items (3.2.25) and (3.2.26), the next items characterize solutions. Here, we use $D_{P'}=2$. Item
(3.7.17) follows from $\Omega_2 = \nu(\nu+D-2)$.

$$\xi = (2+2\nu)\,(\xi'/2) \quad (3.7.13)$$
$$\Omega_2 = \nu(\nu+D_{P'}-2) = \nu(\nu+2-2) = \nu^2 \quad (3.7.14)$$
$$\nu = -1/2 \text{ or } -1 \quad (3.7.15)$$
$$\Omega_2 = \pm S'(S' + D_{P'} - 2) = \pm(S')^2 \quad (3.7.16)$$

- $D = 3 - \Omega_2$, for $\nu = -1$ (3.7.17)
- $D = 5/2 - 2\Omega_2$, for $\nu = -1/2$

Results for D and Ω_2 in item (3.7.17) parallel counterpart results for D and Ω in item (1.4.4).

For #E=2, we can parallel work above, based on substitutions such as those the next items show. Here, $\rightarrow$ denotes replaces.

$$\upsilon \rightarrow \xi \qquad (3.7.18)$$
$$\upsilon' \rightarrow \xi' \qquad (3.7.19)$$
$$t' \rightarrow r \qquad (3.7.20)$$
$$\nabla_{t'}^2 \rightarrow \nabla_r^2 = \nabla^2 \qquad (3.7.21)$$
$$\ddot{\imath} \rightarrow \eta \qquad (3.7.22)$$
$$t \rightarrow x \qquad (3.7.23)$$

The next item restates, in the above coordinates, item (3.2.14).

$$\upsilon \, \Psi(t') = (\upsilon'/2) \left(-\ddot{\imath}^2 \, \nabla_{t'}^2 + \ddot{\imath}^{-2} t'^2 \right) \Psi(t') \qquad (3.7.24)$$

The next item provides an analog to item (3.2.7), which states $0 = \text{Œ}$. Here, we use notation that recognizes that the relevant wave function is a function of both QE-like coordinates and QP-like coordinates. Here, we do not show angular coordinates.

$$0 = \upsilon \, \Psi(r,t') - \xi \, \Psi(r,t') \qquad (3.7.25)$$

The next item defines a symbol with dimensions of velocity.

$$v = \eta \, / \, \ddot{\imath} \qquad (3.7.26)$$

The next items provide characteristics we posit for an edge solution. Here, δ denotes a Dirac delta function. Here, t and t' correlate with items (3.7.23) and (3.7.20), respectively.

$$\Psi \propto \delta(x - vt) \qquad (3.7.27)$$
$$\Psi \propto \delta(r^2 - v^2 t'^2) \qquad (3.7.28)$$
$$\upsilon' = \xi' \qquad (3.7.29)$$

The next item pertains.

$$(\upsilon'/2) \, (\ddot{\imath}^{-2} t^2) - (\xi'/2) \, (\eta^{-2} x^2) = 0 \qquad (3.7.30)$$

The next items show relationships involving operators E and P corresponding to, respectively, energy and momentum for 1-dimensional motion.

$$E^2 \propto (\upsilon'/2) \left(-\ddot{\imath}^2 (\partial^2/\partial t^2) \right) \qquad (3.7.31)$$
$$v^2 P^2 \propto (\xi'/2) \left(-\eta^2 (\partial^2/\partial x^2) \right) \qquad (3.7.32)$$
$$E^2 - v^2 P^2 \propto (\upsilon'/2) \left(-\ddot{\imath}^2 (\partial^2/\partial t^2) \right) - (\xi'/2) \left(-\eta^2 (\partial^2/\partial x^2) \right) \qquad (3.7.33)$$
$$\propto -(\upsilon'/2) \left(\Omega_2(\text{QE-like}) \right) + (\xi'/2) \left(\Omega_2(\text{QP-like}) \right)$$
$$= (\xi'/2) \left(\Omega_2(\text{QP-like}) - \Omega_2(\text{QE-like}) \right)$$

We make interpretations based on the next item.

$$v = c \qquad (3.7.34)$$

For each of $D=D_{P'}=2$ and $D=D_{E'}=2$, the next table shows the only edge solution. Here, $v = -1$. For these edge solutions, $D+2v=0$, for each of QE-like and QP-like.

Table 3.7.1   Edge solutions for 2 dimensions

S'	Ω_2	D	D+2v	
				(3.7.35)
1	1	2	0	(3.7.36)

The next table shows some inside solutions for $v=-1$. Here, 2S' is an even integer. Here, $D+2v = 3-\Omega_2 = (S')^2+1$. [Item (3.7.17)]

Table 3.7.2   v=−1 inside solutions for 2 dimensions

S'	Ω_2	D	D+2v	
				(3.7.37)
0	0	3	1	(3.7.38)
1	−1	4	2	(3.7.39)
2	−4	7	5	(3.7.40)
3	−9	12	10	(3.7.41)
4	−16	19	17	(3.7.42)
5	−25	28	26	(3.7.43)

The next table shows some inside solutions for $v=-1/2$. Here, 2S' is an odd integer. Here, $D+2v = 3/2 - 2\Omega_2$. [Item (3.7.17)]

Table 3.7.3   v=−1/2 inside solutions for 2 dimensions

S'	Ω_2	D	D+2v	
				(3.7.44)
1/2	1/4	2	1	(3.7.45)
1/2	−1/4	3	2	(3.7.46)
3/2	−9/4	7	6	(3.7.47)
5/2	−25/4	15	14	(3.7.48)
7/2	−49/4	27	26	(3.7.49)
9/2	−81/4	43	42	(3.7.50)

~ ~ ~

Fields associated with non-zero-mass elementary bosons correlate with $v=-1$. People might think of $v=-1$ as being 1 dimension inside from a $v=-3/2$ edge.

Fields for elementary fermions correlate with $v=-1/2$. People might think of $v=-1/2$ as being 2 dimensions from a $v=-3/2$ edge. For elementary fermions, we consider parallels to work above, but with 2 dimensions of QE-like partial differential equations correlating with motion and with 2 dimensions of QP-like partial differential equations correlating with motion. [Table 3.4.11] The next item pertains.

Gss.3.7.1   A merger of 2 sets (1 QE-like and 1 QP-like), each of 2 operators, (3.7.57) correlates with a standard representation for $E^2-c^2P^2$ that people use based on the Dirac equation and the Klein-Gordon equation. Operators act on 4-component spinors. People can represent aspects of the operators via gamma matrices.

For #E=2=#P elementary Dirac fermions (which have S=1/2), there remain 1 more dimension for QE-like oscillators and 1 more dimension for QP-like oscillators. For each of the QE-like case and the QP-like case, item (3.2.23) might pertain. We think that the next items pertain.

- Masses for S=1/2 elementary Dirac fermions correlate with exponential functions   (3.7.58)
  - People might consider that remarks following item (3.2.39) correlate with this notion
  - People might consider that Table 3.4.21 correlates with this notion
- Variables for the exponential-like functions may correlate with ...   (3.7.59)
  - Generation
    - People might consider the relevant variable here to be QP-like
    - In Table 3.4.21, mass increases as generation increases
  - Charge for M″=0 particles
    - People might consider the relevant variable here to be QE-like
    - For M″=0 elements in Table 3.4.21, mass decreases as the magnitude of charge increases

~ ~ ~

The next item pertains. For Majorana neutrinos, each of oscillator P2L and P2R is active and each of oscillator E2R and E2L is not active.

Gss.3.7.2   For Majorana neutrinos, wave-equations involving $E^2-c^2P^2$ involve   (3.7.60)
               2-component spinors.

~ ~ ~

The next table describes numbers of components for spinors.

Table 3.7.4   Number of components for spinors for spin-S bosons and fermions

Particle type	Number of components	
Boson	2S+1	(3.7.62)
Dirac fermion	2(2S+1)	(3.7.63)
Majorana fermion	2S+1	(3.7.64)

(3.7.61)

~ ~ ~

For 3Q fermions, the next items pertain.

- For wave-equation operators, people can consider that each of the P0, P2L, and P2R oscillators and 1 of the 2 P4L and P4R oscillators correlates with operators   (3.7.65)
  - This contrasts with, for 1Q and 1L fermions, a list that just includes the P0 oscillator and 1 of the 2 P2L and P2R oscillators
- Consequently, 3Q fermions do not have property-changing interactions with W-family bosons   (3.7.66)

~ ~ ~

The next table pertains. We use (regarding an elementary particle) the term interacts directly with to mean has an interaction vertex with.

Table 3.7.5   Some restrictions on interactions between bosons and fermions

    Gss.3.7.3   Z bosons, the 0H0 boson, and (2S)00 bosons do not directly   (3.7.67)
          participate in interactions that change properties (such as charge) of
          fermions.
      • We assume this, by extension of Table 3.4.11
    Gss.3.7.4   3Q fermions do not interact directly with 22G2& (photons).   (3.7.68)
      • We assume this, by extension of item (3.7.66)

~ ~ ~

Perhaps, for all elementary particles, the next table pertains. Here, m denotes the mass people correlate with the particle.

Table 3.7.6   Definition of sign'(Ω)

    • $E^2 - c^2P^2 = $ sign'(Ω) $|m^2|$ c^4, in which   (3.7.69)
      • sign'$(>0) = 1$
      • sign'$(0) = \pm 1$
        • sign'$(0) = +1$, for the Higgs boson
      • sign'$(<0) = -1$

The mode 0H0 correlates with sign'$(0) = +1$. The next table shows the mode correlating with sign'$(0) = -1$. The mode cannot be excited (for the same reason that, for 22G2&, the P0 oscillator cannot be excited).

Table 3.7.7   The solution 0H0'

E	E	E	E	E	E	E	P	P	P	P	P	P	P	P	P	Particle or	(3.7.70)
6R	6L	4R	4L	2R	2L	0	0	2L	2R	4L	4R	6L	6R	8L	8R	Mode	(3.7.71)
					−1	−1										0H0'	(3.7.72)

People might express concerns regarding sign'$(<0) = -1$. For example, to what extent does this correlate with phenomena in which stuff or information can travel backward in time?

The next table pertains.

Table 3.7.8   Discussion points regarding $\Omega<0$ (or, sign'$(<0) = -1$) behavior

    • Phenomena correlating with sign'$(<0) = -1$ are quantum mechanical   (3.7.73)
      • People might say that classical-physics notions of trajectories do not fully
        pertain

- After the era of inflation, phenomena correlating with sign'(<0) = −1 are   (3.7.74)
  confined to limited ranges of space-time coordinates
  - For situations correlating with experiments, people might say that
    lengths R_0 characterize QP-like bounds
    - For example, people might point to the sizes of mesons and baryons
  - For the relevant bosons (O- and Y-families), people might say that life
    spans are short and correlate with QE-like bounds
- During the era of inflation, phenomena correlating with sign'(<0) = −1 are   (3.7.75)
  confined to limited ranges of space-time coordinates
  - Compared to (say) the universe during and after the initial formation of
    nuclei, people might say that QP-like inflation-era bounds are small
  - Compared to (say) the universe during and after the initial formation of
    nuclei, QE-like inflation-era bounds are small

We know of nothing in observations or in this paper that we think rules out item (3.7.69). We discuss implications below. [For example, discussion preceding and following Table 4.2.3]

~ ~ ~

The next item pertains.

- Table 3.7.1 and Table 3.7.2 contribute to work in Section 4.2 regarding   (3.7.76)
  ratios of masses of non-zero-mass elementary particles

~ ~ ~

Ref.3.7.1   Wolfram Alpha, computational knowledge engine, Wolfram Alpha LLC,
             http://mathworld.wolfram.com/Laplacian.html.

Section 3.8   Mathematics related to uncertainty

Abs.3.8.1   We show a way to derive an uncertainty principle for non-zero-mass elementary particles.
  ▪   Math we use provides a way to study uncertainty related to base states.

The next item follows from item (3.2.14).

$$\xi = (\xi'/2) (\eta^2 <p_r^2> + \eta^{-2} <r^2>), \text{ in which}$$
$$<j> \text{ denotes the expected value of } j \qquad (3.8.1)$$

The next item follows from item (3.2.25).

$$D + 2\nu = \eta^2 <p_r^2> + \eta^{-2} <r^2> \qquad (3.8.2)$$

$\eta^{-2} <r^2>$ does not depend on η. Similarly, $\eta^2 <p_r^2>$ does not depend on η. The 2 terms contribute equally. The next items pertain for $\xi' \neq 0$.

- $<p_r^2> \times <r^2>$ does not vary with changes in η^2, for $\eta^2 > 0$ (3.8.3)
- For edge solutions, people might consider that $<r^2> = 0$ (3.8.4)

The next items sketch deriving an uncertainty-like equation. We think that this equation pertains for at least W- and H-family bosons.

- Follow steps correlating with items (3.7.1) through (3.7.28)   (3.8.5)
- Set $\upsilon' = -\xi'$ (3.8.6)
  - Compare with item (3.7.29)
- Note the following (compare with item (3.7.33))   (3.8.7)
  - $\ddot{\imath}^2 E^2 + \ddot{\imath}^{-2} t^2 + \eta^2 c^2 P^2 + \eta^{-2} c^2 x^2 \propto$ (components from 2-dimensional QE-like terms and components from 2-dimensional QP-like terms)

Part 4   Invariant and relative properties

Section 4.0   Introduction and summary

Abs.4.0.1   We predict masses for some elementary particles.
- This work provides opportunities to detect or rule out some possible elementary particles.

In Part 4, we predict masses {or mass-math eigenvalues} for neutrinos, of which 2 would be Majorana fermions (~0.125 eV/c^2 and ~0.058 eV/c^2) and 1 would be a Dirac fermion (~0.058 eV/c^2). We predict masses (from 80.4 GeV/c^2 to 113.7 GeV/c^2) and lower bounds for cluster-production threshold energies (from ≥241 GeV to ≥569 GeV) for O-family elementary particles. We discuss the applicability of concepts of invariant and relative.

Traditional physics theory discusses concepts of invariant properties and relative properties. People might say that concepts related to special relativity correlate with usefulness for distinguishing invariant from relative.

So far in this paper, we have built from quantum concepts and provided models correlating with phenomena. People might say that the Part 1 catalog of elementary particles is an invariant list. People might say that Part 2 observations generally lie beyond needs to make theoretical distinctions between invariant concepts and relative concepts. People might say that Part 3 features invariant concepts.

In Part 4, we continue to work on invariant concepts and we start to emphasize relative concepts.

Section 4.1   Q-mass, C-mass, and rest mass

Abs.4.1.1   We define C-mass (roughly, classical-physics mass) and Q-mass (roughly, quantum-physics mass).
- This work provides a step toward math for modeling general-relativistic kinematics in a way that accounts for zero-mass particles having energy.

Work in Section 3.7 leads to a quantum mechanical expression for $E^2 - c^2P^2$ for elementary particles. The next table describes concepts regarding mass.

Table 4.1.1   C-mass and Q-mass, for an elementary particle or other object
- m   (4.1.1)
  - Mass (or, energy/c^2) correlating with the m in $E^2 - c^2P^2 = \pm|m^2|c^4$
    - Here, E and P are quantum operators
      - People might correlate E with energy
      - People might correlate P with momentum
- C-mass   (4.1.2)
  - Mass (or, energy/c^2) correlating (roughly) with classical physics
  - Here, m does not include contributions from the cloud of virtual particles (for example, virtual fermion/anti-fermion pairs) that correlate with the particle or other object

- Q-mass   (4.1.3)
  - Mass (or, energy/c^2) correlating with interactions via gravity
  - Here, m includes contributions from the cloud of virtual particles (for example, virtual fermion/anti-fermion pairs) that correlate with the particle or other object

We anticipate finding significance for each of C-mass and Q-mass. For example, we anticipate that photons have C-mass=0. [Section 4.5] People say that gravity alters the paths of photons. Quantum mechanically, gravitons would not interact directly with photons. Quantum mechanically, people discuss non-zero amplitudes for elementary-fermion particle-antiparticle pairs associated with a photon. Gravitons interact with elementary fermions. Thus, quantum mechanical mechanisms exist for gravitons to alter paths of photons. The energy associated with a photon is a relative (not an invariant) phenomenon. People might say that gravity alters photon trajectories via concepts related to Q-mass. (And, people might consider that use of concepts of curved space time and geodesic trajectories may not be needed to discuss photon trajectories (and possibly any trajectories). [Section 6.5])

Based on item (3.5.26), we equate PR44G4 with Q-mass.

The next item pertains.

Table 4.1.2   Q-mass, C-mass, rest mass, invariant, and relative
- Experiments and observations measure Q-mass   (4.1.4)
- For a particles or other object with non-zero C-mass, ...   (4.1.5)
  - People might correlate the term relative (to a specific reference frame) with Q-mass
  - People might use the term rest mess to denote the minimum Q-mass that an observer can measure
    - People might say that the particle or object is at rest in a frame of reference correlating with such a minimum measurement of Q-mass
    - People might correlate the term invariant with that rest mass
- For a particle (that is, a G- and Y-family elementary particle) with zero C-   (4.1.6)
  mass, ...
  - People might correlate the term relative with Q-mass
    - There is no minimum for possible observations for Q-mass
    - No observer sees the particle as being at rest
  - Traditionally, people use the term rest mass to refer to C-mass
    - People might correlate the term invariant with that rest mass

Section 4.2   Q-masses and charges for W-, H-, and O-family particles

Abs.4.2.1   Masses for O-family bosons are approximately 80.4, 91.2, 105.3, and 113.7 GeV/c^2. Threshold energies for creating minimum numbers of O-family bosons may be at least ≥241, ≥274, ≥402, ≥526, and ≥569 GeV.
  - This work points to possible opportunities for experimental research.
Abs.4.2.2   The 2O2 boson has charge $-(1/3)|q_e|$ and the 2O1 boson has charge $+(1/3)|q_e|$.
  - This work provides some understanding of quantum aspects related to fractional charge.

The next items provide a formula that correlates with Q-masses for W-family elementary bosons (2W%) and the H-family elementary boson (0H0). Here, m(j) denotes the Q-mass of elementary boson j.

$$m`` = m(Z\ boson) / 3 \qquad (4.2.1)$$
$$(m(2W\%))^2 \approx (m``)^2 \times f(2W\%) \qquad (4.2.2)$$
$$(m(0H0))^2 \approx (m``)^2 \times f(0H0) \qquad (4.2.3)$$
$$f(2W0) = 9 \qquad (4.2.4)$$
$$f(2W1) = f(2W2) \approx 7 \qquad (4.2.5)$$
$$f(0H0) = 17 \qquad (4.2.6)$$

The next table compares calculated and experimental masses. [Ref.4.2.1, Ref.4.2.2, and Ref.4.2.3] Here, jX% denotes a symbol in the symbol column.

Table 4.2.1   Calculated and experimental masses for W- and H-family bosons   (4.2.7)

Particle	Symbol	f(jX%)	Calculated mass (GeV/c²)	Experimental mass (GeV/c²)	
Z	2W0	9	91.188	91.1876±0.0021	(4.2.8)
W	2W1, 2W2	7	80.420	80.385±0.015	(4.2.9)
Higgs	0H0	17	125.325	125.3 ± 0.4 (stat) ± 0.5 (sys) [Ref.4.2.2]	(4.2.10)
				126.0 ± 0.4 (stat) ± 0.4 (sys) [Ref.4.2.3]	

Here, we equate values of D+2v (related to Ω_2 [Section 3.7]) with contributions to the squares of masses. The next items correlate with f(jX%)-column results that, respectively, items (4.2.8), (4.2.9), and (4.2.10) show.

- The square of the Q-mass of a Z boson correlates with the sum of   (4.2.11)
  - −1 = minus a QE-like instance of item (3.7.38)
  - +10 = plus a QP-like instance of item (3.7.41)
- The square of the Q-mass of a W boson correlates with the sum of   (4.2.12)
  - −2 = minus a QE-like instance of item (3.7.39)
  - −1 = minus a QE-like instance of item (3.7.38)
  - +10 = plus a QP-like instance of item (3.7.41)
- The square of the Q-mass of a Higgs boson correlates with the sum of (1 term)   (4.2.13)
  - +17 = plus a QP-like instance of item (3.7.42)

The next items pertain regarding item (4.2.9).

- Generally, in nature, observed masses of zero-charged particles are larger than observed masses for similar charged particles   (4.2.14)
  - For example, the mass of a neutron is larger than the mass of a proton
  - For example, the mass of a neutral pion is larger than the mass of a charged pion
- For W bosons, the difference between calculated Q-mass and experimental mass correlates with differences in virtual-particle clouds associated with zero-charged particles and virtual-particle clouds with charged particles   (4.2.15)

For the W-, H-, and O-families, we do not attempt (in this paper) to estimate C-masses.

~ ~ ~

In this subsection, we discuss charges for O-family bosons.

Table 1.5.1 shows representations for ground states of 2O% bosons. Based on Œ=0, exciting an N(Ej) from 0 to 1 requires exciting an N(Pk) from 0 to 1. We know of no preferred value of Pk.

The next item pertains to the 2O1 particle and shows an amplitude for such an excitement for oscillator E1. Here, we use notation that denotes | N(E2R) , N(E2L) , N(E0) , N(P0) , N(P2L) , N(P2R) >.

$$(1/3)^{1/2} \, | \, \# \, , 1 \, , \# \, , 1 \, , 0 \, , 0 > \qquad\qquad (4.2.16)$$
$$+ \, (1/3)^{1/2} \, | \, \# \, , 1 \, , \# \, , 0 \, , 1 \, , 0 >$$
$$+ \, (1/3)^{1/2} \, | \, \# \, , 1 \, , \# \, , 0 \, , 0 \, , 1 >$$

The charge for a 2O N(Ej)=0 item in the Table 1.5.1 correlates with the charge associated with the Pj oscillator. The charges associated with the other 2 values of Pk are not relevant.

The next item pertains. Here, as above, Q' denotes charge divided by $|q_e|$.

$$Q'(2O1) \qquad\qquad\qquad (4.2.17)$$
$$= ((1/3)^{1/2} < \# \, , 1 \, , \# \, , 0 \, , 1 \, , 0 \, |) \, ((1/3)^{1/2} \, | \, \# \, , 1 \, , \# \, , 0 \, , 1 \, , 0 >) \, Q'(2W1) + 0$$
$$= (1/3) \, Q'(2W1)$$

No 4W elementary particles exist. 3Q elementary particles do not interact directly with photons. [Table 3.7.5] The next table pertains.

Table 4.2.2   Charges for O-family bosons
  • Q'(2Oj) = (1/3) Q'(2Wj), for j = 3, 2, and 1   (4.2.18)
  • Q'(4Oj) = 0, for j = 5, 4, 3, 2, and 1   (4.2.19)

~ ~ ~

The next table shows how patterns pertaining to W- and H-family Q-masses might pertain to O-family masses. Here, f parallels the f in item (4.2.7) in Table 4.2.1. The f column shows the sum of numbers in the columns to the left of the f column. For the Q-mass column, we use the formula Q-mass = (m(Z boson)/3) $|f|^{1/2}$. The minimum number of bosons (min. numb. of bosons) column echoes concepts that O-family particles must be created in multiples [Item (3.4.137)]. The threshold column shows minimum estimated energies needed to create the particles. For the O-family, we compute numbers by multiplying the Q-mass by c^2 and by the minimum number of bosons. In effect, we assume that zero contribution is needed regarding binding energy. This assumption may not be correct. (For quarks, the binding energy is non-zero. For example, the mass of a proton exceeds the sum of the masses of the 3 quarks contained in the proton.) We explain the patterns (which we use for the leftmost 5 columns in these items) below. [Items including and following item (4.2.30)] The f=−17 possibility [Item (4.2.24)] correlates with OHO'. [Table 3.7.7] We treat the f=−17 possibility as a math solution that does not correlate with an elementary particle.

Table 4.2.3   Possible Q-masses and threshold energies for O-family bosons   (4.2.20)

E4R, E4L	E2R, E2L	E0, P0	P2L, R2R	P4L, P4R	f	Q-mass (GeV/c²)	Min. numb. of bosons	Threshold (GeV)	Bosons (or math solution)	
		−1	10		9	91.2	1	91.2	2W0	(4.2.21)
	−2	−1	10		7	80.4	1	80.4	2W1, 2W2	(4.2.22)
		17			17	125.3	1	125.3	OHO	(4.2.23)

E4R, E4L	E2R, E2L	E0, P0	P2L, R2R	P4L, P4R	f	Q-mass (GeV/c^2)	Min. numb. of bosons	Threshold (GeV)	Bosons (or math solution)	
		-17			-17				(0H0')	(4.2.24)
	-10	1			-9	91.2	3	≥ 274	200	(4.2.25)
	-10	1	2		-7	80.4	3	≥ 241	202, 201	(4.2.26)
-5	-10	1			-14	113.7	5	≥ 569	400	(4.2.27)
-5	-10	1	2		-12	105.3	5	≥ 526	402, 401	(4.2.28)
-5	-10	1	2	5	-7	80.4	5	≥ 402	404, 403	(4.2.29)

(4.2.20)

The next items describe patterns.

- Integers (shown in the E4R, E4L column through the P4L, P4R column) reflect Ω_2-related values of D+2v for the 0≤S'≤4 inside solutions [Table 3.7.2] (4.2.30)
- The S'=1 edge solution [Table 3.7.1] pertains wherever Table 4.2.3 shows a blank in 1 of the table's 5 leftmost columns (4.2.31)
- For the O-family, ... (4.2.32)
  - A QE-like instance of inside/S'=2 correlates with the E4R, E4L pair
  - A QE-like instance of inside/S'=3 correlates with the E2R, E2L pair
  - A QP-like instance of inside/S'=0 correlates with the E0, P0 pair
  - A QP-like instance of inside/S'=1 correlates with the P2L, P2R pair
  - A QP-like instance of edge/S'=2 correlates with the P4L, P4R pair
- Items correlating with the W-family or the H-family reflect work above in Table 4.2.1 and Table 3.7.7 (4.2.33)

Also, we note that, for a kX% particle, values in the oscillator-related columns in Table 4.2.3 are zero for oscillators ElR, ElL, PlL, and PlR for which l>k.

~ ~ ~

Ref.4.2.1   J. Beringer et al. (Particle Data Group), *PR D86*, 010001 (2012) and 2013 partial update for the 2014 edition (URL: http://pdg.lbl.gov). (http://pdg.lbl.gov/2013/tables/rpp2013-sum-gauge-higgs-bosons.pdf)
Ref.4.2.2   CMS collaboration (2012). "Observation of a new boson at a mass of 125 GeV with the CMS experiment at the LHC". *Physics Letters B* 716 (1): 30–61. arXiv:1207.7235. Bibcode:2012PhLB..716...30C. doi:10.1016/j.physletb.2012.08.021.
Ref.4.2.3   ATLAS collaboration (2012). "Observation of a New Particle in the Search for the Standard Model Higgs Boson with the ATLAS Detector at the LHC". *Physics Letters B* 716 (1): 1–29. arXiv:1207.7214. Bibcode:2012PhLB..716....1A. doi:10.1016/j.physletb.2012.08.020.

Section 4.3   A possible correlation between the Higgs-boson mass, ℏ, c, and G_N

Abs.4.3.1   An arithmetic combination of ℏ, c, and G_N approximates (within less than 8 parts in 100) the mass of a Higgs boson.

- This work possibly relates the masses of W-, H- and O-family bosons to ℏ, c, and G_N.

Work in this paper features basis states. People might say that no traditional uncertainty applies regarding kinematics for basis states. (People might say the same thing regarding plane-wave solutions people use within traditional quantum mechanics.) In traditional plane-wave kinematics, the uncertainty principle and its use of the constant ℏ arise regarding the superposition of basis-state wave functions. Presumably, the same happens here.

We explore that possibility that our work produces, from a quantum basis and for 1 basis state, the notion that ℏ pertains to uncertainties.

The next item pertains.

Gss.4.3.1   The confluence of at least 2 G-family series {out of 3 e-family series - (4.3.1)
the series of N(P0)=−1 bosons (44G4& and 22G2&), the series of #E=0
bosons (84G2468&, 64G246&, 42G24&, and 22G2&), and the j2G2& series
(82G2&, 62G2&, 42G2&, and 22G2&)} at 22G2& correlates with the role of
ℏ in the uncertainty principle.

Items including and following item (3.8.5) provide a possible basis for an uncertainty calculation regarding non-zero-mass elementary bosons.

Based on observations, the next item pertains. Here, m(0H0) denotes the mass of a Higgs boson.

$$(m(0H0))^2 / ℏ ≈ 0.85 (c^5 / G_N) \qquad (4.3.2)$$

Perhaps, people will find that an expression for mass squared divided by ℏ has appeal, based on item (3.8.7). (People might say that item (3.8.7) represents that a square of energy scales with an uncertainty.) Perhaps, people will find that masses for W-, H-, and O-family particles correlate with the values of ℏ, c, and G_N.

The next item pertains.

Tbd.4.3.1   To what extent is there significance that, in formulas pertaining to a (4.3.3)
series of lengths [Table 3.5.1], G_N^{-1} correlates with c^0 and G_N^{+1} correlates
with $ℏ^0$?

Section 4.4   Q-masses and charges for L-family and spin-1/2 Q-family particles

Abs.4.4.1   A formula approximates masses for 6 quarks and 3 charged leptons.
- This work may help substantiate the use of math models that people might say include a negative range for a radial coordinate.

Abs.4.4.2   If sibling states exist, the formula approximates masses for siblings of 3 charged leptons.
- This work may help people direct searches for sibling states.

Abs.4.4.3   Possibly, masses or mass-math eigenvalues (in units of eV/c^2) and types for the baryonic-matter neutrinos are ~0.125 Majorana, ~0.058 Majorana, and ~0.058 Dirac.
- This work may point to numbers correlating with masses or mass-math eigenvalues for neutrinos.

The next items apply for lepton particles and for quark particles. Here, L denotes a principal quantum number for spin-like systems that correlate with $-r_E \leq r \leq 0^+$. [Item (3.2.39)] Here, M denotes a secondary quantum number. We use results from Table 3.2.2.

- For each 1 of 3 generations, the following number of relevant solutions   (4.4.1)
  correlates with each of S=1/2 with Ω=+3/4 and S=1/2 with Ω=−3/4
  - $2(2S+1) = 4$

Gss.4.4.1   For the L-family, combinations of the 4 solutions correspond to 2 of   (4.4.2)
the 3 possible members of an L=1 set (the M=0 member does not pertain) and to the 1 member (M=0) of each of 2 L=0 sets.

Gss.4.4.2   For the Q-family, for S=1/2, combinations of the 4 solutions   (4.4.3)
correspond to 4 of the 5 members of an L=2 set (the M=0 member does not pertain).

We interpret Table 3.4.21 as correlating with and supporting items (4.4.2) and (4.4.3). In Table 3.4.21, M" and M' are integer indices. The table shows an orderly array of approximate particle masses for charged leptons and quarks. Here, M correlates with (but does not necessarily equal) M'.

The next items correlate with item (4.4.2), item (4.4.3), and Table 3.4.21. Here, we consider all 3 generations.

- For each of the rows for which M" = 0, 2, or 3, the following pertain   (4.4.4)
  - The row includes an instance characterized by L=1
    - M' = −3, 0, and +3 characterize this instance
    - A lepton and anti-lepton pair correlate with M' = −3 and M' = +3
    - No particle correlates with M' = 0 for that row
    - A corresponding L = 0, M = 0 row has M" ≤ −3
- For each of the rows for which M" = 0, 1, or 2, the following pertain   (4.4.5)
  - The row includes an instance characterized by L = 2
    - M' = −2, −1, 0, +1, and +2 characterize this instance
    - A quark and anti-quark pair correlate with M' = −2 and M' = +2
    - A quark and anti-quark pair correlate with M' = −1 and M' = +1
    - No particle correlates with M' ~ 0

- For each of the rows for which M" = 0 or 2, the following pertain   (4.4.6)
  - The row includes an instance characterized by L = 3
  - M' = −3, −2, −1, 0, +1, +2, and +3 characterize this instance
  - A lepton and anti-lepton pair correlate with M' = −3 and M' = +3
  - A quark and anti-quark pair correlate with M' = −2 and M' = +2
  - A quark and anti-quark pair correlate with M' = −1 and M' = +1
  - No particle correlates with M' = 0

Edge solutions correlate with Q- and L-family particles. Above, we discuss possibilities that considering $\Psi(r)$, with r<0, makes sense regarding such particles. [Item (3.2.39)] We note possibilities for roles for exponential and trigonometric functions.

The next item defines notation regarding Table 3.4.21.

- m(M",M') denotes a calculated Q-mass correlating with a position in a table   (4.4.7)
  that includes positions for S=1/2 charged elementary fermions

The next item pertains to trends in Table 3.4.21.

Gss.4.4.3   For charged leptons (either M'=−3 or M'=+3), people can benefit by   (4.4.8)
correlating the range −1≤M"≤3 with an L=2 system.

The next item pertains. For the L=2 system with fixed M' and varying M", −2≤M"−1≤+2. (Regarding the next item, we think that no solution with d(0)=−d(2) appeals. We think such a solution would correspond to a trigonometric function that is anti-symmetric with respect to M"−1. The next item features an expression that corresponds to a symmetric trigonometric function.)

Gss.4.4.4   For the L=2 system (with fixed M' and varying M") that includes   (4.4.9)
charged leptons and for some number ζ", $m(M",−3) \propto e^{M"\zeta"}(1+d(M"))$, in
which −1≤M"≤3, d(0)=d(2), and d(−1)=d(1)=d(3)=0.

The next items show results for M'=−3 charged leptons and related values of M". Here, we use experimental masses for the electron and muon to calculate ζ". Then, we use an experimentally acceptable calculated mass for the tauon [Item (2.6.2)] to calculate m`. Then, we calculate d(0).

$$m(M",−3) = m` \times \exp((M"+1)\zeta") \times (1+d(M")) \quad (4.4.10)$$
$$\zeta" = (1/2) \log(m_{muon}/m_e) \approx 2.665799 \quad (4.4.11)$$
$$m` = m_{tauon} / \exp(4\zeta") \approx 4.155987\times10^{-2} \text{ MeV/c}^2 \quad (4.4.12)$$
$$1 + d(0) = m_e / (m` \exp(\zeta")) \quad (4.4.13)$$
$$d(2) = d(0) \approx −0.144926 \quad (4.4.14)$$
$$d(−1) = d(1) = d(3) = 0 \quad (4.4.15)$$

The next item provides a calculated number that does not directly correlate with a particle.

$$m(1,−3) \approx 8.59326 \text{ MeV/c}^2 \quad (4.4.16)$$

The next items correlate with a possible role for trigonometric functions.

$$d(M") \approx d" \times (1/2) \times (\cos(M"\pi) + 1) \quad (4.4.17)$$
$$d" = d(0) = d(2) \approx −0.144926 \quad (4.4.18)$$

We guess that the next item provides a useful set of m(M'',0). Here, α denotes the fine-structure constant. No known particles correlate with the combination M''≥0 and M'=0.

$$m(M'', 0) \approx m(M'', -3) \times \exp((1/4) \log(1/\alpha)\ 3(1+M'')) \qquad (4.4.19)$$

The next items show numbers.

$$m(0, 0) \approx 2.05 \times 10^1 \text{ MeV/c}^2 \qquad (4.4.20)$$
$$m(1, 0) \approx 1.38 \times 10^4 \text{ MeV/c}^2 \qquad (4.4.21)$$
$$m(2, 0) \approx 6.79 \times 10^6 \text{ MeV/c}^2 \qquad (4.4.22)$$
$$m(3, 0) \approx 4.57 \times 10^9 \text{ MeV/c}^2 \qquad (4.4.23)$$

The next items finish defining the formula correlating with approximate masses for quarks and charged leptons. We specify d(M'',M') adequate to fit known experimental results. (For each M'', we could use a trigonometric function to specify d(M'', M').)

- $m(M'', M') \approx m(M'', 0) \times \exp((1/4) \log(\alpha)\ (1+M'')\ |M'|) \times (1+d(M'',M'))$ (4.4.24)
- $d(M'', \pm 2) = -d(M'', \pm 1) \approx d`(-d``)^{M''}$, for 0≤M''≤2 and 2≥|M'|≥1   (4.4.25)
- $d(M'', M') = 0$ otherwise
- $d` \approx 0.2$ (4.4.26)
- $d`` \approx 0.4$ (4.4.27)

The next items pertain.

$$-d(0, -1) = d(0, -2) \sim 0.2 \qquad (4.4.28)$$
$$-d(1, -1) = d(1, -2) \sim -0.08 \qquad (4.4.29)$$
$$-d(2, -1) = d(2, -2) \sim 0.032 \qquad (4.4.30)$$
$$d(M'', M') = 0, \text{ otherwise within } 0 \leq M'' \leq 3, 3 \geq |M'| \geq 0 \qquad (4.4.31)$$

The next 2 tables compare calculated numbers with experimental numbers. The tables show masses in units of MeV/c². For each particle, the bottom number (calc) comes from our calculations.

The next table compares calculated masses with experimental masses for charged leptons. [Ref.10.5.3]

Table 4.4.1   Experimental and calculated Q-masses for charged leptons

M''		M' −3	...	...	
					(4.4.32)
					(4.4.33)
0	exp	0.510998928±0.000000011			(4.4.34)
0	calc	0.510998928 MeV/c²			(4.4.35)
1	exp				(4.4.36)
1	calc				(4.4.37)
2	exp	105.6583715±0.0000035			(4.4.38)
2	calc	105.6583715			(4.4.39)
3	exp	1776.82±0.16			(4.4.40)
3	calc	1776.81			(4.4.41)

The next table compares calculated masses with experimental masses for quarks. [Ref.4.4.1] Except regarding M''=2 quarks, the upper number (exp) comes from experiments. For M''=2 quarks, Ref.4.4.1 provides two possible ranges for quark masses. The upper range is based on mass-independent subtraction scheme(s) (MS). For M"=2, M'=−2, the first mass is the MS running mass and the second mass is the 1S mass. For M"=2, M'=−1, the first mass is labeled MS from cross-section measurements and the second mass is from direct measurements.

Table 4.4.2   Experimental and calculated Q-masses for quarks

		M' ...	−2	−1	
M"					(4.4.42)
					(4.4.43)
0	exp		$2.3^{+0.7}_{-0.5}$	$4.8^{+0.7}_{-0.3}$	(4.4.44)
0	calc		2.10 MeV/c^2	4.79	(4.4.45)
1	exp		95 ± 5	$(1.275\pm0.025)\times10^3$	(4.4.46)
1	calc		92.5	1.272×10^3	(4.4.47)
2	exp		$(4.18\pm0.03)\times10^3$	$(160^{+5}_{-4})\times10^3$	(4.4.48)
			$(4.65\pm0.03)\times10^3$	$(173.5\pm0.6\pm0.8)\times10^3$	
2	calc		4.367×10^3	164.1×10^3	(4.4.49)
3	exp				(4.4.50)
3	calc				(4.4.51)

Possibly, m(M",M') correlates with each of the 9 masses for quarks and charged leptons. People might say that we use the following numbers to fit the data.

$$m`, \zeta", d(0), \alpha, d`, \text{ and } d`` \tag{4.4.52}$$

The next item correlates with work above. Here, we assume d(−1,±3)=d(−1,0)=0.

$$m(-1,-3) = m(-1,0) = m(-1,+3) = m` \tag{4.4.53}$$

~ ~ ~

The next item pertains. [Item (4.4.24)]

$$m(M", M') \sim m(M", 0) \times ((\alpha)^{-1/4(1+M")})^{-|M'|}, \text{ for } 3\geq|M'|\geq1 \tag{4.4.54}$$

The next table shows items that possibly correlate values of |M'| with Ω_2-related values of D+2v. [Table 3.7.3]

Table 4.4.3   Possible relationships between |M'| and Ω_2-related values of D+2v

|M'|		
1	D+2v for Ω_2=1/4	(4.4.55)
2	D+2v for Ω_2=−1/4	(4.4.56)
3	(D+2v for Ω_2=1/4) + (D+2v for Ω_2=−1/4)	(4.4.57)
		(4.4.58)

The next item pertains.

Tbd.4.4.1   Determine the extent to which the 2 S'=1/2 rows of Table 3.7.3 correlate with item (4.4.10). $\qquad$ (4.4.59)

~ ~ ~

We point to an involvement of square roots of masses.

For charged leptons, the next item pertains. [Items (4.4.10) and (4.4.11)] The formula involves integer powers of the square roots of 2 masses.

$$m(M'', M') \approx m` \times ((m_{muon} / m_e)^{(1/2)})^{(M''+1)} \times (1+d(M'')) \qquad (4.4.60)$$

The next item shows the Koide formula. People correlate this formula with aspects of the Standard Model.

$$(m_e + m_{muon} + m_{tauon}) / (m_e^{1/2} + m_{muon}^{1/2} + m_{tauon}^{1/2})^2 \approx 2/3 \qquad (4.4.61)$$

The next item reflects nominal numbers we use, including the calculated value for m_{tauon}. The uncertainty-range may not be accurate.

$$(m_e + m_{muon} + m_{tauon}) / (m_e^{1/2} + m_{muon}^{1/2} + m_{tauon}^{1/2})^2 \approx 0.66666(\sim 3) \qquad (4.4.62)$$

The next item pertains.

Tbd.4.4.2   To what extent might people find significance in the relationship   (4.4.63)
$(\tan^{-1}(1+d'') - 2^{-1/2}) / 2^{-1/2} \sim 4.6 \times 10^{-4}$?

~ ~ ~

The next table correlates with items (3.5.3) and (3.5.6). In particular, in item (4.4.66), k=5 for tauons.

Table 4.4.4   Some G-family forces and their relative strengths   (4.4.64)

Force	Channels	Electron relative vertex strength per channel	Electron k for β^{-k}	Tauon relative vertex strength per channel	Tauon k for β^{-k}	
22G2&	4	$1 = \beta^{-0}$	0	$1 = \beta^{-0}$	0	(4.4.65)
44G4&	3	β^{-6}	6	β^{-5}	5	(4.4.66)

The next table includes information about the masses of a muon.

Table 4.4.5   Some information about a pattern pertaining to masses of charged leptons

Particle	Symbol for mass $m(M'',M')$	mass/m_e	M''	
electron	$m(0,-3) = m_e$	$\beta^{0/3}$	0	(4.4.68)
-			1	(4.4.69)
muon	$m(2,-3)$	$\sim \beta^{(2/3)\cdot(1-0.02)}$ or $\sim 0.9 \cdot \beta^{2/3}$	2	(4.4.70)
tauon	$m(3,-3)$	$\beta^{3/3}$	3	(4.4.71)

(4.4.67)

~ ~ ~

We try to extend methods above to predict neutrino masses (or mass-math eigenvalues). Item (4.4.2) suggests 2 L=0 sets, each with 1 member.

People interpret observations as implying that the sum of the neutrino masses does not exceed 0.28 eV/c^2. [Ref.4.4.2 and Ref.4.4.3] Other observations correlate with the existence of exactly 3 types of light neutrinos. [Ref.4.4.4]

The next table shows numbers based on formulas above. Possibly, some m(M",M') correlate with masses for zero-charge fermions. Possibly, numbers in the table have relevance. Each might correlate with 1 of the 2 L=0 sets.

Table 4.4.6   Some numbers that might be related to neutrino Q-masses

M"	M'	m(M",M')	(4.4.72)
−6	−3	0.058 eV/c^2	(4.4.73)
−3	0	0.125	(4.4.74)

In Section 2.7, we correlate 3 solutions with the term generations and a number, 3, of generations. For neutrinos, perhaps another term, such as flavors, might be more appropriate than generations.

Possibly, the next table correlates with 3 neutrino masses or mass-math eigenvalues. For the M'=−3 item in the table above, perhaps a neutrino correlates with a state that is a (an equal) linear combination of an M'=−3 state and an M'=+3 state. Based on symmetries, possibly the M'=−3 item correlates with each of a Dirac state (1 linear combination, possibly proportional to the difference of the M'=−3 state and the M'=+3 state) and a Majorana state (1 linear combination, possibly proportional to the sum of the M'=−3 state and the M'=+3 state). We think that a linear combination proportional to a difference of 2 similar fermion states correlates with the symmetry for each of the 2 participating states. We think that a linear combination proportional to a sum of 2 similar fermion states correlates with a different symmetry. Possibly, the M'=0 state correlates with a Majorana state. [Table 3.4.19]

Table 4.4.7   Possible masses (in units of eV/c^2) or mass-math eigenvalues for neutrinos

M"	|M'|	m(M",±M')	Combination (of M'=−3 and M'=+3 states)	Symmetry	Type	(4.4.75)
−6	−3	0.058 eV/c^2	Anti-symmetric	SU(5)	Dirac fermion	(4.4.76)
−6	−3	0.058	Symmetric	SU(7) or SU(3)×SU(2)×U(1)	Majorana fermion	(4.4.77)
−3	0	0.125	(Not applicable)	SU(7)	Majorana fermion	(4.4.78)

The next table provides a parallel to Table 3.4.22. The zero-charge 1L column pertains to neutrinos and possible siblings of neutrinos. In that column, D denotes Dirac fermion and M denotes Majorana fermion. The notation BMS'(j') denotes possibilities for 1≤j'≤5 siblings of the BMS'(j'=0) neutrinos that Table 4.4.7 lists. For each value of j', the column shows 3 fermion solutions. For j' = 1, 3, and 5, we include a Dirac fermion solution (and a Majorana fermion with the same M"), even though the corresponding BMS6 solution does not correspond to a charged lepton. Because SU(5) symmetry provides for just 3 Dirac zero-charge leptons per peer, we think that 3 solutions (j' = 1, 3, and 5) do not correlate with elementary particles. We denote those 3 solutions by (D), not by D. For the charged 1L column j for BMS6(j), we hypothetically extend the range of j to include 2 negative values. For the charged 1L column for n` for U(1):n`, we hypothetically extend the use of n`. For the charged 1L column for k for SU2(k), we hypothetically extend the range of k to include a value of 0. We do not extend (compared to Table 3.4.22) the concept columns.

Thomas.J.Buckholtz@gmail.com   Copyright (c) 2014 Thomas J. Buckholtz   http://ThomasJBuckholtz.wordpress.com

Table 4.4.8   Sets of zero-charge lepton solutions and charged-lepton solutions for the peer MP(0)   (4.4.79)

M"	zero-charge 1L: j', type(s) j' for BMS'(j') D or M for type	charged 1L: j for BMS6(j)	charged 1L: concept (n`=1)	charged 1L: concept (n`=2)	charged 1L: n` for U(1):n`	charged 1L: k for SU(2):k	(4.4.80)
−6	0; D, M	−2			1, 2	0	(4.4.81)
−5		−2			2	0	(4.4.82)
−4		−2			2	0	(4.4.83)
−3	0; M	−1, −2			1, 2	0	(4.4.84)
−2	1; (D), M	−1			1	0	(4.4.85)
−1		−1			1	0	(4.4.86)
0		−1, 0	electron		1, 2	0, 1	(4.4.87)
1	1; M					1	(4.4.88)
2	2; D, M	0	muon		1	1	(4.4.89)
3		0, 1	tauon	a-tauon	1, 2	1	(4.4.90)
4						1	(4.4.91)
5	2; M	1		positron	2	1	(4.4.92)
6	3; (D), M	1, 2	muon	a-muon	1, 2	1, 2	(4.4.93)
7						2	(4.4.94)
8		2	tauon		1	2	(4.4.95)
9	3; M	2, 3	electron	positron	1, 2	2	(4.4.96)
10	4; D, M					2	(4.4.97)
11		3		a-muon	2	2	(4.4.98)
12		3, 4	tauon	a-tauon	1, 2	2, 3	(4.4.99)
13	4; M					3	(4.4.100)
14	5; (D), M	4	electron		1	3	(4.4.101)
15		4, 5	muon	a-muon	1, 2	3	(4.4.102)
16						3	(4.4.103)
17	5; M	5		a-tauon	2	3	(4.4.104)
18	-	5		positron	2	3	(4.4.105)

If siblings of neutrino solutions correlate with elementary particles, work above correlates with totals (including the 3 light neutrinos) of 12 Majorana zero-charge leptons and 3 Dirac zero-charge leptons.

The next table provides masses for possible BMS'(j'=1) zero-charge particles. We derive the numbers via the same technique we used for Table 4.4.7.

Table 4.4.9   Possible masses or mass-math eigenvalues for BMS'(j'=1) zero-charge particles   (4.4.106)

M"	|M'|	$m(M", \pm M')$	Combination (of M'=−3 and M'=+3 states)	Symmetry	Type	
−2	−3	2.5×10^3 eV/c^2	Symmetric	SU(7) or SU(3)×SU(2)×U(1)	Majorana fermion	(4.4.107)
+1	0	14 GeV/c^2	(Not applicable)	SU(7)	Majorana fermion	(4.4.108)

The next table provides masses for possible BMS'(j'=2) zero-charge particles. We derive the numbers via the same technique we used for Table 4.4.7.

Table 4.4.10 Possible masses or mass-math eigenvalues for BMS'(j'=2) zero-charge particles (4.4.109)

M"	|M'|	m(M",±M')	Combination (of M'=−3 and M'=+3 states)	Symmetry	Type	
+2	−3	1.06×10^2 MeV/c^2	Anti-symmetric	SU(5)	Dirac fermion	(4.4.110)
+2	−3	1.06×10^2 MeV/c^2	Symmetric	SU(7) or SU(3)×SU(2)×U(1)	Majorana fermion	(4.4.111)
+5	0	1.5×10^{12} GeV/c^2	(Not applicable)	SU(7)	Majorana fermion	(4.4.112)

The next table provides masses for possible BMS'(j'=3) zero-charge particles. We derive the numbers via the same technique we used for Table 4.4.7.

Table 4.4.11 Possible masses or mass-math eigenvalues for BMS'(j'=3) zero-charge particles (4.4.113)

M"	|M'|	m(M",±M')	Combination (of M'=−3 and M'=+3 states)	Symmetry	Type	
+6	−3	4.5×10^3 GeV/c^2	Symmetric	SU(7) or SU(3)×SU(2)×U(1)	Majorana fermion	(4.4.114)
+9	0	1.7×10^{23} GeV/c^2	(Not applicable)	SU(7)	Majorana fermion	(4.4.115)

Experiments place lower limits on masses of stable neutral heavy lepton masses at 45 GeV/c^2 for Dirac fermions and 39 GeV/c^2 for Majorana fermions. [Ref.4.4.5] Based on these limits BMS'(j'=1) states either do not exist or can decay. Based on these limits BMS'(j'=2) M"=+2 states either do not exist or can decay.

We think that the limit of 3 types of light neutrinos does not pertain regarding BMS'(j'>0) states. The next items pertain.

Tbd.4.4.3   How might observations detect effects of BMS'(j'>0) zero-charge leptons? (4.4.116)

Tbd.4.4.4   Try to create BMS'(j'>0) zero-charge leptons. (4.4.117)

Tbd.4.4.5   Effect, detect, infer, or rule out (to some confidence level) the populating of the BMS'(1) M"=−2 zero-charge state in situations for which temperatures correlate with at least (or for which energies exceed) 2.5×10^3 eV. (4.4.118)

Tbd.4.4.6   Effect, detect, infer, or rule out (to some confidence level) the populating of the BMS'(1) M"=+1 zero-charge state in situations for which temperatures correlate with at least (or for which energies exceed) 14 GeV. (4.4.119)

Tbd.4.4.7   Effect, detect, infer, or rule out (to some confidence level) the populating of BMS'(2) M"=+2 zero-charge states in situations for which temperatures correlate with at least (or for which energies exceed) 1.06×10^2 MeV. (4.4.120)

Tbd.4.4.8   Effect, detect, infer, or rule out (to some confidence level) the populating of the BMS'(3) M"=+6 zero-charge state in situations for which energies exceed (or for which temperatures correlate with at least) 4.5×10^3 GeV. (4.4.121)

Tbd.4.4.9   To what extent to do solutions based on mathematics herein correlate with masses (or mass-math eigenvalues) for zero-charge leptons? (4.4.122)

Tbd.4.4.10 Mathematically, to what extent does the lack of M"=1 charged   (4.4.123)
            leptons correlate with the existence of an M"=1 Majorana zero-charge
            lepton solution?

Tbd.4.4.11 Mathematically, to what extent does the muon mass not matching   (4.4.124)
            $\beta^{2/3}m_e$ [Item (4.4.70)] correlate with the existence of at least 1 of the M"=2
            zero-charge lepton solutions?

~ ~ ~

In Section 8.2, we conjecture about possible C-masses for electrons, tauons, and neutrinos.

~ ~ ~

We discuss the extent to which BMS6(j>0) particles and BMS'(j'>0) particles interact directly with photons.

In item (3.4.205), we find that 1 set of photons spans MP(0). BMS6(j>0) leptons would have non-zero charge. BMS'(j'>0) leptons would have zero charge. The next table pertains.

Table 4.4.12 Interactions between photons and BMS6(j>0) or BMS'(j'>0)

- MP(0) photons interact directly with BMS6(j>0) particles   (4.4.125)
- MP(0) photons do not interact directly with BMS'(j'>0) particles   (4.4.126)

~ ~ ~

The next item provides possible relationships regarding masses and charges for the lowest-mass charged lepton in each sibling. Here, m(3j,−3) denotes the mass of the lowest-mass charged lepton in sibling BMS6(j). Here, m(3j,−3) matches, via 3j=M" and −3=M', notation we use in Table 3.4.21 and Table 3.4.22. Here, $\beta'=m_{tauon}/m_e$. (Per channel, a gravitational (44G4&) interaction between 2 M"=18 charged leptons would have the same magnitude as an electromagnetic (22G2&) interaction between the same particles. [Item (3.5.9)]) Here, we recognize that solutions BMS6(1), BMS6(3), and BMS6(5) do not correlate with elementary particles. [Table 3.4.20]

Gss.4.4.5   Regarding masses and charges for lepton siblings BMS3(2j), for 2j =   (4.4.127)
             0, 2, or 4, m(3×2j,−3) = $(\beta')^{2j}$×m(0,−3) and |Q'(3×2j,−3)| = |Q'(0,−3)|.

The next items pertain.

- The least-mass non-zero-charge BMS6(j>0) lepton would have mass   (4.4.128)
  m(6,−3) ~ $4.5×10^3$ GeV/c²

Tbd.4.4.12 Effect, detect, infer, or rule out (to some confidence level) the   (4.4.129)
            populating of BMS6(j>0) non-zero-charge lepton states.

~ ~ ~

As yet, we do not have theory that estimates masses for 3Q fermions. Possibly, a mass-related number related to 3Q is 3 times the corresponding mass-related number for 1Q fermions. Here, we have in mind interpreting R_0 as a ratio of spin to mass. [Item (3.5.16)] The ratio of spins for 3Q to spins for 1Q is 3 (=(3/2)/(1/2)). Perhaps the mass-related number is m(0,0). We do not further discuss masses for the possible S=3/2 fermions.

Like quarks, S=3/2 fermions would not range freely in today's universe. The next items pertain.

Tbd.4.4.13 Try to detect or create 3Q fermions.   (4.4.130)

Tbd.4.4.14 Try to detect or create compound particles that include 3Q fermions. (4.4.131)

Tbd.4.4.15 To the extent S=3/2 elementary fermions exist, determine masses   (4.4.132)
for S=3/2 elementary fermions.

Tbd.4.4.16 To the extent S=3/2 elementary fermions exist, determine masses   (4.4.133)
for nucleon-like clusters of S=3/2 elementary fermions.

Tbd.4.4.17 To the extent S=3/2 elementary fermions exist, determine masses   (4.4.134)
for nuclei-like clusters of nucleon-like clusters of S=3/2 elementary
fermions.

~ ~ ~

Ref.4.4.1   J. Beringer et al. (Particle Data Group), *Phys. Rev. D86*, 010001 (2012).

Ref.4.4.2   S. Thomas, F. Abdalla, and O. Lahav, Upper Bound of 0.28 eV on the Neutrino Masses from the
Largest Photometric Redshift Survey, *Phys. Rev. Lett. 105*, 031301, 2010.
(http://arxiv.org/abs/0911.5291)

Ref.4.4.3   A. Melchiorri, Constraints on Neutrino Physics from Planck, European Space Agency,
http://www.rssd.esa.int/SA/PLANCK/docs/eslab47/Session06_CMB_Cosmology_and_Fundamental_
Physics/47ESLAB_April_04_17_30_Melchiorri.pdf.

Ref.4.4.4   K.A. Olive *et al.* (Particle Data Group), Chin. Phys. C38, 090001 (2014) (URL:
http://pdg.lbl.gov) - "Number of Neutrino Types"

Ref.4.4.5   K.A. Olive *et al.* (Particle Data Group), Chin. Phys. C38, 090001 (2014) (URL:
http://pdg.lbl.gov) - "Leptons"

Section 4.5   C-masses of G- and Y-family bosons

Abs.4.5.1   For G- and Y-family elementary particles, C-mass = 0.

- This work bridges from our work to traditional concepts that photons and gluons have no
mass.

The next table shows the number of Ω_2(QE-like) contributions and the number of Ω_2(QP-like)
contributions that pertain to C-mass for some G-family members. Each Ω_2(QE-like) or Ω_2(QP-like)
contribution contributes to C-mass an amount proportional to D+2v=0. [Table 3.7.1]

Table 4.5.1   Contributions to C-mass, for G-family #E=0 elementary particles

Boson	Number of Ω_2(QE-like)	Number of Ω_2(QP-like)	(4.5.1)
22G2&	0	1	(4.5.2)
42G24&	0	2	(4.5.3)
64G246&	0	3	(4.5.4)
84G2468&	0	4	(4.5.5)

The next item pertains to each #E=0 member of the G-family.

$$C\text{-mass} = 0 \qquad (4.5.6)$$

Item (4.5.6) pertains for all G-family elementary particles.

~ ~ ~

Within a choice of 1 MP(n) (for n = 0, 1, or ...), the next table depicts the 1 relevant instance of gravitons.

Table 4.5.2   Graviton (44G4&) modes and their ground states, within 1 MP(n)

E 6R	E 6L	E 4R	E 4L	E 2R	E 2L	E 0	P 0	P 2L	P 2R	P 4L	P 4R	P 6L	P 6R	P 8L	P 8R	Particle or Mode	
																Particle or Mode	(4.5.7) (4.5.8)
		#	#	0	−1	#	#	0	#							44G4L	(4.5.9)
		#	#	0	−1	#	#	#	0							44G4R	(4.5.10)

Also, the next table shows the other G-family #E=2 modes.

Table 4.5.3   42G2& modes and their ground states, within 1 MP(n)

E 6R	E 6L	E 4R	E 4L	E 2R	E 2L	E 0	P 0	P 2L	P 2R	P 4L	P 4R	P 6L	P 6R	P 8L	P 8R	Particle or Mode	
																Particle or Mode	(4.5.11) (4.5.12)
		#	#	0	−1	0	#	#	#							42G2L	(4.5.13)
		#	#	0	−1	#	0	#	#							42G2R	(4.5.14)

Above [Section 1.1, Section 1.2, and Section 1.3], we equate, for purposes of describing effects of polarizations, P4L for gravitons with P2L for photons. And, for the same purposes, we equate P4R for gravitons with P2R for photons.

Perhaps, people can use a common set of space-time coordinates to describe mathematics pertaining to all the MP(n).

The next item pertains.

> Gss.4.5.1   Quantum mechanically, the 2 sets of coordinates Table 1.1.6   (4.5.15)
> presents suffice as a basis for modeling G-family #E=0 and #E=2
> elementary-particle kinematics across all MP(n).

~ ~ ~

The next table shows the number of Ω_2(QE-like) contributions and the number of Ω_2(QP-like) contributions that pertain for Y-family members. Each contribution correlates with a pair of oscillators. Each Ω_2(QE-like) or Ω_2(QP-like) contribution contributes to C-mass an amount proportional to D+2v=0. [Table 3.7.1]

Table 4.5.4   Contributions to C-mass, for Y-family elementary particles

Bosons	Number of Ω_2(QE-like)	Number of Ω_2(QP-like)	
			(4.5.16)
2Y	1	1	(4.5.17)
4Y	2	2	(4.5.18)

The next item pertains to each member of the Y-family.

$$C\text{-mass} = 0 \qquad (4.5.19)$$

Part 5   Dark energy, dark matter, and sibling-state fermions

Section 5.0   Introduction and summary

Abs.5.0.1   We describe dark-energy stuff, dark matter, and possibilities for detectable sibling states.
- ▪ This work provides opportunities to resolve various physics-theory problems people associate with the term unsolved.

In Part 5, we describe elementary particles that could provide bases for dark matter and for dark-energy stuff. We also describe possible siblings (related to known leptons) that people might call elementary particles.

Section 5.1   Dark energy

Abs.5.1.1   Dark-energy stuff consists of as many as 7 peers of the stuff people associate with the combination of baryonic matter and dark matter.
- ▪ This work provides a candidate explanation for dark-energy stuff.

The next table extends work from Table 3.4.15 and Table 3.4.17.

Table 5.1.1   Terminology related to dark-energy stuff
- We use the term NDEF (an acronym for non-dark-energy fermions) to denote the peer that includes baryonic matter   (5.1.1)
- We use the term DEF (an acronym for dark-energy fermions) to denote the entirety of the 7 peers other than NDEF
- Dark-energy stuff includes 7 peers of NDEF   (5.1.2)
- We number the peers so that MP(0) denotes NDEF [Item (3.4.135)]   (5.1.3)

No G-, W-, Y-, or O-family forces connect MP(0) through MP(3) stuff with MP(4) through MP(7) stuff. The next table pertains.

Table 5.1.2   Sources of possibly observable effects from the existence of peers MP(4) through MP(7)
- Observable effects of MP(4) through MP(7) are limited to...   (5.1.4)
  - Asymmetries early in the big bang
    - For example, perhaps an asymmetry in the distributions of stuff pertains
  - Interactions mediated by Higgs bosons (0H0)

The next table pertains.

Table 5.1.3   Ultimate impact of NDEF-peers on calculations of densities of the universe
- To the extent each of the peers MP(0), MP(1), MP(2), and MP(3) were created adequately similarly and have maintained adequately similar evolutions, observations might eventually lead to (hypothetical future) beings estimating that DEF contributes at least 3/4 of the density of the universe   (5.1.5)

Observations suggest a ratio of ~(2.2):1 for the densities of the universe for DEF and NDEF. [Ref.5.1.1] In Section 8.1, we show why today's observational results do not necessarily need to dovetail with the possibility that there is (today) and has been (for essentially the entire past history of the universe) a ratio of 3:1 for MP(1)+MP(2)+MP(3) to MP(0).

~ ~ ~

Ref.5.1.1   Mark Peplow, Planck telescope peers into primordial universe, *Nature News*, Nature Publishing Group, March 21, 2013. (http://www.nature.com/news/planck-telescope-peers-into-primordial-universe-1.12658)

Section 5.2   Sibling states

Abs.5.2.1   We discuss the possible existence of siblings of leptons.
- Experimental opportunities may exist to verify or rule out sibling states.

This next table summarizes some solutions that may correlate with elementary particles. [Table 3.4.19, Table 3.4.27, and Table 4.4.8] BMS6(1), BMS6(3), and BMS6(5) do not correlate with elementary particles. [For example, Section 3.4]

Table 5.2.1   Candidates for sibling-state fermions
- BMS3(j) for j>0   (5.2.1)
  - These particles would be charged leptons
  - To the extent BMS3(j>0) states exist, they include 2 siblings (BMS6(2) and BMS6(4)) of baryonic-matter charged leptons
- BMS'(j') for j'>0   (5.2.2)
  - These particles would be zero-charge leptons
  - To the extent BMS'(j'>0) states exist, there are 5 siblings (BMS'(1) through BMS'(5)) of baryonic-matter Majorana neutrinos
  - To the extent BMS'(j'>0) states exist, there are 2 siblings (BMS'(2) and BMS'(4)) of baryonic-matter Dirac neutrinos

MP(0) includes only 1 instance of W-family particles. BMS6(j>0) particles could decay into BMS'(0) particles (that is, neutrinos) or into BMS'(j>0) particles (that is, states that people might correlate with the term heavy neutrinos). Per Table 4.4.9 and remarks following that table, BMS'(1) states are not stable. The next item pertains.

Gss.5.2.1   For each j'≥2, BMS'(j') states are unstable.   (5.2.3)

Per item (4.4.127) and other work in Section 4.4, the only BMS6(j>0) or BMS'(j'>0) states with mass less than the mass of a tauon would be the 1 M"=−2 zero-charge particle and the 2 M"=+2 zero-charge particles. The mass of a tauon exceeds the mass of a proton. The next items pertain.

- Except for the 1 M"=−2 zero-charge state in BMS'(1) and the 2 M"=+2 zero-   (5.2.4)
  charge states in BMS'(2), all BMS6(j>0) and BMS'(j'>0) states significantly
  depleted by the time in the evolution of the universe that nucleons formed

- The 1 M"=−2 zero-charge state in BMS'(1) significantly depleted around the   (5.2.5)
  time in the evolution of the universe that people might correlate with a
  temperature correlating with the energy 2.5×10^3 eV
- The 2 M"=+2 zero-charge states in BMS'(2) significantly depleted around   (5.2.6)
  the time in the evolution of the universe that people might correlate with a
  temperature correlating with the energy 1.06×10^2 MeV/c^2
- To the extent BMS6(j>0) and BMS'(j'>0) states exist, such particles may be   (5.2.7)
  created under circumstances in which adequate energy is available (for
  example, in situations for which adequately high temperature pertains)

Items (4.4.118), (4.4.120), and (4.4.121) invite work to effect, detect, infer, or rule out some possible siblings of neutrinos.

Section 5.3   Dark matter

Abs.5.3.1   Dark matter consists of at least 1 of spin-3/2 fermions and siblings of baryonic-matter leptons.
- This work points to possible opportunities to observe or infer properties of dark matter.

The next table describes candidates for DMFSSF (an acronym for dark-matter fermions or sibling-state fermions). [Table 3.4.19, Table 3.4.27, and Table 4.4.8] We note that item (3.4.159) correlates with a lack of siblings for quarks.

Table 5.3.1   Candidates for DMFSSF (or, dark-matter or sibling-state fermions)
- To the extent 3Q elementary fermions exist, DMFSSF includes spin-3/2   (5.3.1)
  elementary fermions
- To the extent j>0 siblings of baryonic-matter charged leptons exist, DMFSSF (5.3.2)
  includes 2 siblings (BMS3(1)=BMS6(2) and BMS3(2)=BMS6(4)) of some
  baryonic-matter leptons
- To the extent j'>0 siblings of baryonic-matter neutrinos exist, DMFSSF   (5.3.3)
  includes 5 siblings (BMS'(1) through BMS'(5)) of baryonic-matter Majorana
  neutrinos and 2 siblings (BMS'(2) and BMS'(4)) of baryonic-matter Dirac
  neutrinos

We discuss possible properties of siblings in Section 4.4. We allude to the topic of properties for 3Q fermions and related compound particles in items (4.4.132), (4.4.133), and (4.4.134).

~ ~ ~

The next item pertains. [Table 3.7.5 and Table 4.4.12]

- Within DMFSSF, only BMS6(j>0) particles interact directly with 22G2&   (5.3.4)
  (photons)

~ ~ ~

We know of no direct interaction that would transform fermions between 3Q states and 1Q states.

~ ~ ~

The next table summarizes thoughts regarding DMFSSF. Each entry in the aspect column begins with the word may.

Table 5.3.2   Dark-matter fermions and/or sibling-state fermions

Aspect	BMS6(j>0) charged leptons	BMS'(j'>0) zero-charge leptons	3Q fermions and 3Q-based compound particles	(5.3.5)
• May exist	Yes	Yes	Yes	(5.3.6)
• May be created (under appropriate circumstances)	Yes	Yes	Yes	(5.3.7)
• May depopulate in circumstances of adequately low temperature	Yes	Yes	No	(5.3.8)
• May have depopulated early in the history of the universe	Yes	Yes	No	(5.3.9)
• May currently contribute to what people might interpret as dark matter	Yes	Yes	Yes	(5.3.10)
• May have contributed previously to what people might interpret as dark matter	Yes	Yes	Yes	(5.3.11)

Part 6   New models for phenomena for which people say traditional theories correlate

Section 6.0   Introduction and summary

Abs.6.0.1   We explore relationships between models in this work and traditional models correlating with spin; helicity, chirality, and handedness; conservation laws; some aspects of photonics; and effects people model via general relativity.
- This work provides a quantum basis for explaining phenomena people correlate with general relativity.
- This work provides a basis for avoiding concepts of curved space times.

In Part 6, we offer seemingly straightforward ways to discuss spin and to discuss helicity, chirality, and/or handedness. Also, we correlate with our quantum-based approach various aspects of conservation of energy, momentum, and angular momentum; photonics; and general relativity.

Section 6.1   Spin

Abs.6.1.1   We offer seemingly a straightforward way to model spin.
- We offer math correlating with spin that is more invariant or compelling than some math in traditional approaches.

People may find less than adequately compelling some traditional discussions of spin for elementary particles. For example, people might find questionable some concepts related to apparently physical rotation within point particles.

People might say that we offer an invariant approach. We do not discuss possible physical rotation. People might say we do not discuss possible rotation with respect to space-time coordinates.

We correlate spin with the number of QP-like oscillators that correlate with a particle.

~ ~ ~

We use representations in which S correlates with QP-like aspects and perhaps not much with QE-like aspects. People might say that this basis for spin appears to be incompatible with special relativity.

However, people might say that our treatment of kinematics (such as that, for each elementary particle, $E^2-c^2P^2=$constant) points to adequate compatibility with special relativity. (We address the applicability of special relativity in Section 8.5.) Thus, people might say that using a mostly QP-like treatment of spin does not introduce problems.

Section 6.2   Helicity, chirality, and/or handedness

Abs.6.2.1   We offer seemingly straightforward ways to model helicity, chirality, and/or handedness.
- We offer math correlating with helicity, chirality, and/or handedness that is more invariant or compelling than some math in traditional approaches.

People may find less than adequately simple some traditional treatments of helicity, chirality, and/or handedness for elementary particles. Perhaps people would correlate some difficulty with people trying to envision physical rotation within point particles. Perhaps people would correlate some difficulty with people trying to envision an apparent difference in property for an elementary particle, depending on whether the particle is moving toward or away from an observer.

People might say we offer an invariant approach. We emphasize the signs of the charges of elementary boson particles with which elementary fermions interact.

~ ~ ~

The next table provides symbols that people might say reinforce remarks above. Here, for example, 2W2L' refers to a fermion's losing a unit of positive charge by absorbing a W⁺ boson. (Compare with item (1.6.24) for a particle that can absorb a W⁺ and with item (1.7.22) for notation regarding erasing a unit of color charge.)

Table 6.2.1   Alternative notation for charged W-famiIy elementary particles

Particle	Symbol	Symbol related to absorption	Symbol related to emission	
				(6.2.1)
W⁺ boson	2W1	2W2L'	2W2L	(6.2.2)
W⁻ boson	2W2	2W2R'	2W2R	(6.2.3)

Section 6.3   Conservation of energy, momentum, and angular momentum

Abs.6.3.1   We show examples of conservation of energy, momentum, and angular momentum.
- These examples show the possibility for deriving the three conservation laws for all interactions.

We discuss thought experiments regarding interactions, each of which destroys a photon. We imagine that the photon is emitted by an object. We discuss happenings when an elementary particle or other absorber absorbs the photon.

Without loss of generality, we discuss left circularly polarized photons. (Similar work pertains to right circularly polarized photons.) The next item shows the first excited state for a left circularly polarized photon. [Table 1.1.4]

Table 6.3.1   A once excited left circularly polarized photon

E	E	E	E	E	E	P	P	P	P	P	P	P	P	P	Particle or	
6R	6L	4R	4L	2R	2L	0	0	2L	2R	4L	4R	6L	6R	8L	8R	Mode	(6.3.2)
					1	−1	1	#								22G2L	(6.3.3)

~ ~ ~

We imagine an absorption for which no change of energy or momentum takes place. For example, we imagine that each of the emitter and absorber is an atom in a crystal. The emission is the de-excitement of a non-valence electron in 1 atom. The absorption is the excitement of a non-valence electron in another atom. Here, people might say that, for the absorbing atom, the internal angular momentum changes. Here, people might say that, for the photon, absorbing correlates with a change to N(E0) and a change to N(P2L). The next item shows the final state for the photon.

Table 6.3.2   A once excited photon, after de-excitation via an orbital-angular-momentum interaction

E	E	E	E	E	E	P	P	P	P	P	P	P	P	P	P	
6R	6L	4R	4L	2R	2L	0	0	2L	2R	4L	4R	6L	6R	8L	8R	(6.3.4) (6.3.5)
				0	-1	0	0									(6.3.6)

We think we can generalize. The next item pertains.

> Gss.6.3.1   Our approach exhibits conservation of angular momentum.   (6.3.7)

~ ~ ~

We imagine that an absorption takes place in a way that, in the rest frame of the emitter, no angular momentum changes take place. People might say that electrostatic attraction between originally stationary charged objects provides an example. Here, changes in motion would be limited to effects of electrostatic attraction or repulsion. The photon transfers energy via oscillator E0 and momentum via oscillator P0. The next table shows the result of the absorber absorbing the photon. Here, people might say that transmission of energy to the absorber correlates with changes to N(Ej) values. Here, people might say that transmission of momentum to the absorber correlates with changes to N(Pj) values. Regarding the overall interaction, people might say the neither the emitter nor the absorber changed invariant properties.

Table 6.3.3   A once excited photon, after de-excitation via an electrostatic interaction

E	E	E	E	E	E	P	P	P	P	P	P	P	P	P	P	
6R	6L	4R	4L	2R	2L	0	0	2L	2R	4L	4R	6L	6R	8L	8R	(6.3.8) (6.3.9)
				0	0	0	0	0	0							(6.3.10)

We think we can generalize. The next item pertains.

> Gss.6.3.2   Our approach exhibits conservation of energy and conservation of   (6.3.11)
> momentum.

The next item pertains.

> Gss.6.3.3   Within the physics of 1 instance of MP(n), people can consider that   (6.3.12)
> the W-family ground state and the 22G2& ground state have similarities.

Section 6.4   Photonics - refraction, absorption, and reflection

Abs.6.4.1   We explore roles of relative phenomena and invariant phenomena for photons.
  ▪   Parallels, for gravitons, for this work might lead to new ways to test for gravitons.

We discuss thought experiments regarding interaction of a photon with an object. The next table shows relevant aspects. For discussion, we use a reference frame in which the object is at rest. We assume that, outside the object, the photon travels in a vacuum. The table shows an incoming path for the photon. The table indicates, for when the photon is at a certain point on the path, the cloud of virtual particles we associate with the photon. We use the symbol + to denote part of the cloud. Per discussion above, that cloud correlates with a non-zero Q-mass for the photon. For these examples, we assume the photon is a singly-excited left circularly polarized photon. (The assumption of direction of polarization is arbitrary. The assumption of excitation number is a convenience.)

Table 6.4.1   Diagram of an incoming photon and an object

		Photon path	(6.4.1)
	↙		(6.4.2)
+ ↙		Cloud of	(6.4.3)
↙ +		virtual particles	(6.4.4)
↙			(6.4.5)
↙			(6.4.6)
-- -- -- -- -- -- -- -- -- -- -- --		Top surface	(6.4.7)
		Object	(6.4.8)
-- -- -- -- -- -- -- -- -- -- -- --		Bottom surface	(6.4.9)
			(6.4.10)

People might say that some of the vocabulary we use in this section correlates too much with classical-physics vocabulary (for example, trajectory) and not enough with even wave-mechanics (for example, plane wave). Perhaps, however, the points we make do not depend on such distinctions.

~ ~ ~

We discuss refraction. We assume that each surface is adequately smooth. We assume that the surfaces parallel enough other.

While the photon approaches the top surface, the object interacts more with some parts of the cloud than with others. We symbolize the former parts of the cloud as being to the lower-right of the path. We symbolize the latter parts of the cloud as being to the upper left of the path. Based on this asymmetry, the photon path bends downward. After the photon has significantly passed the top surface, interaction between the cloud and the object stops significantly bending the trajectory.

When the photon exits the object, the lower-right part of the cloud leaves the object before the upper-left part of the cloud leaves. Then, the trajectory parallels (but does not match a straight-line continuation of) the original trajectory.

Upon exit, aside from the displacement of the trajectory, the photon is in a state similar to the state it occupied approaching the object. For the photon, Q-mass is conserved. Also, the invariant properties remain unchanged.

People might take a broader look at angular momentum. People might consider the point at which the trajectory is halfway through the object. Relative to that point, after exiting the object, the photon has clockwise positive (as seen from the perspective of looking straight into the diagram) angular momentum, whereas, before entering the object, the photon has clockwise negative angular moment. People might say that the object experiences a torque during the photon's transiting the object.

~ ~ ~

We discuss absorption.

Here, the photon trajectory ends within the object. The object absorbs energy, momentum, and angular momentum correlating with Q-mass for the photon. For the photon, the invariant representation changes to correlate with de-excitation.

~ ~ ~

We discuss reflection.

For now, we assume the original photon trajectory is normal to the top surface of the object. (Regarding Table 6.4.1, the trajectory is straight downward.)

Here, the departing photon carries the same energy as the incoming photon. Here, the interaction deposits with the object downward momentum correlating with twice the Q-mass of the incoming photon.

If the interaction deposits no angular momentum with the object, the departing photon has 1 unit of right circular polarization. The invariant representation for the photon changes.

If the interaction deposits angular momentum with the object, the departing photon has 1 unit of left circular polarization. The invariant representation for the photon does not change.

We assume people can generalize to cases for which the original photon trajectory is oblique.

~ ~ ~

The next items pertain. And, results might lead to new concepts for tests regarding gravitons. [Item (9.3.16)]

Tbd.6.4.1 To what extent might work in photonics benefit from people distinguishing phenomena related to relative (photon) quantities from phenomena related to invariant (photon) quantities?	(6.4.11)

Section 6.5   General relativity

Abs.6.5.1   We suggest quantum bases for phenomena people might correlate with general relativity.
- This work provides opportunities to determine the extent to which quantum mechanics provides models that underlie aspects of general relativity.

We consider some thought experiments regarding particles or other objects moving under the influence of 44G4& (gravitons) associated with an object we term the hub. We set up the thought experiments so that there is a reference frame in which the hub remains stationary. We imagine 2 focal points, point-1 and point-2, positioned to be equidistant from the hub, such that the hub bisects a straight line between the 2 focal points. We imagine a plane that includes the 3 points. All the phenomena we discuss occur in this plane. We imagine ellipse-1, which has focal points at point-1 and the hub. We imagine ellipse-2, which has focal points at the hub and point-2. We imagine that 4 identical objects orbit the hub. 2 objects have orbits that track ellipse-1. These 2 objects orbit in opposite directions. The other 2 objects have orbits that track ellipse-2. These 2 objects orbit in opposite directions. In the frame of reference in which the hub is stationary, all 4 objects arrive simultaneously at the perihelia for their respective orbits. We ignore collisions between objects. For these circumstances, momentum and angular momentum balance adequately so that that the hub remains stationary in the frame of reference. We consider, for each of some examples, models regarding 1 orbiting object.

We consider an example in which the hub is a non-rotating black hole, the orbiting object is a photon and point-1 and point-2 are at the center of the hub. Here, the orbit for the photon is a circle. Here, the orbit occurs within the event horizon for the black hole. The photon has C-mass=0 and Q-mass≠0. Gravitons (44G4&) associated with the hub do not interact directly with the photon. The non-zero Q-mass correlates with a cloud of virtual particles, including virtual pairs of fermions, associated with the photon. Gravitons (44G4&) associated with the hub interact with the virtual fermions in that cloud. Gravitational attraction, based on non-zero Q-mass, keeps the photon in its orbit. Each of the photon's observed energy and observed momentum is proportional to the photon's Q-mass. The next items pertain.

- In the chosen frame of reference, the expression (Q-mass)c^2 describes the perceived energy associated with the photon (6.5.1)
- In the chosen frame of reference, the expression (Q-mass)c describes the perceived magnitude of momentum associated with the photon (6.5.2)

We consider another example in which the orbiting object is a photon and orbit occurs within the event horizon of a hub that is a black hole. This time, point-1 and point-2 do not coincide with the center of the hub. The orbit is an ellipse that is not a circle. Here, the photon's Q-mass varies, depending on where the photon is in its orbit. The photon's Q-mass is greatest at the orbit's perihelion. The photon's Q-mass is least at the orbit's aphelion. The next items pertain.

- In any frame of reference, the expression (Q-mass)c^2 (as measured in that frame of reference) describes the perceived energy associated with a photon (6.5.3)
- In any frame of reference, the expression (Q-mass)c (as measured in that frame of reference) describes the perceived magnitude of momentum associated with a photon (6.5.4)

In Section 7.3 and Section 7.4, we discuss other such examples.

~ ~ ~

In Section 2.2, we discuss the concept that #E=0 G-family members set a large-scale context of zero-curvature or little-curvature. Above, we show possible similarities between dark-energy stuff and baryonic matter plus dark matter. People might think that such points to difficulties in modeling general-relativistic phenomena based on notions of curved space time. For example, how can people develop models (based on space-time coordinates) that embrace more than 1 instance of MP(n)? Also, people might say that models based on general relativity may not be adequately invariant - for example with respect to Q-mass. The next item pertains.

Gss.6.5.1   All G-family physics can be discussed based on the flat space-time coordinates Table 1.1.6 shows. (6.5.5)

The next item pertains.

- People can represent gravitation (and each of the other G-family forces) as additive (across sources), based on 4-vector constructs (6.5.6)
  - People need to consider the spans of such forces
    - Such spans correlate with the extent to which forces mediate interactions between elementary particles in different peers
  - People need to base forces on the respective Q-property (for example, regarding 44G4&, Q-mass) for each of the interacting objects
  - For #E=0 forces and #E=2 forces, the 4-vector constructs are similar to the 4-vector vector potential for 22G2&

~ ~ ~

A rotating mass produces a 44G4& analog to the 22E2& magnetic field produced by a rotating object that has a net charge and for which charges reside in adequately fixed places in the object. By analogy with electromagnetism, for 2 currents of mass moving parallel (not antiparallel) to each other, the 44G4& analogy to the 22E2& magnetic field produces forces that attract the 2 currents of mass toward each other.

Part 7   Models that may correlate with phenomena

Section 7.0   Introduction and summary

Abs.7.0.1   We provide correlations between this work and various phenomena.
- This work solves or points to how to solve various problems involving the extent to which traditional theories correlate with observed phenomena.
- This work points to opportunities or observational and theoretical research.

In Part 7, we provide quantum-based models correlating with planetary perihelia shifts and other phenomena people traditionally model via general relativity. Also, we discuss inflation, asymptotic freedom, the lack of observing free quarks or gluons, black-hole thermal radiation, a force that helps hold galaxies together, the spacecraft flyby anomaly, the extent to which neutrinos are Dirac fermions or Majorana fermions, vacuum zero-point energy and the cosmological constant, and potential correlations with nuclear physics.

We acknowledge that material, including a list of unsolved problems in physics, from Wikipedia proved useful regarding this part and other parts.

Section 7.1   Inflation

Abs.7.1.1   Observations that people term inflation correlate with kinematics of Q-family fermions.
- This work suggests an era in which quarks or spin-3/2 quark-like particles ranged freely or ranged within a fermion sea involving many such particles.

People interpret observations as correlating a rapid increase in size of the universe between approximately 10^{-36} seconds after the beginning of the big bang and approximately 10^{-33} to 10^{-32} seconds after the beginning of the big bang. People use the term inflation to denote relevant phenomena. Some people say that (in some sense) space time expanded at faster than the speed of light.

People say that a quark-and-gluon plasma existed until the formation of individual nucleons. People estimate that nucleons formed between approximately 10^{-6} seconds and 1 second after the big bang.

In Section 3.7, we use space-time coordinates to develop operators E^2 and P^2 that people might say correlate with, respectively the square of energy and the square of momentum. For elementary particles that have non-zero C-mass and non-zero spin, the next item pertains. [Table 3.7.6 and Table 3.7.8]

$$\bullet \quad E^2 - c^2P^2 = \text{sign}'(\Omega) \times (mc^2)^2 \qquad\qquad (7.1.1)$$

For the Q-family and for the O-family, $\text{sign}'(\Omega)=-1$. People might say that freely-ranging Q-family particles or O-family particles correlate with motion at speeds greater than c, the speed of light. The next item pertains. Here, we allow the interpretation that free can mean motion within a sea of fermions that includes many fermions. (Perhaps, people could draw parallels between such a phenomena and phenomena related to electron-based superconductivity.)

Gss.7.1.1   Inflation correlates with conditions such that at least 1 of the   (7.1.2)
        following featured free particles - 1Q and 3Q.

During inflation, the stuff that people have observed (that is, the currently observable or inferable universe) collected (from a large volume) within a small volume. After that era, FE1 forces drove expansion. (Presumably, FEI forces also drove expansion before and during inflation.)

(To the extent people make statements based on assumptions that speeds are limited to the speed of light, people might say that the stuff that collected in the small volume expanded rapidly from a much smaller volume. We disagree with assigning speeds of less than the speed of light. Indeed, perhaps the classical-physics concept of speed does not pertain. We disagree with the notion that the stuff came exclusively from a much smaller volume.)

We recognize that (during inflation) instances of the same types of particles (that collected within the stuff that people can observe (or infer)) may have left the realm that correlates with stuff people have observed.

Section 7.2   Asymptotic freedom and lack of measurements of free quarks or gluons

Abs.7.2.1   We discuss lack of observation of free quarks and gluons and we discuss the notion of asymptotic freedom.
- This work supports the notion that observing free quarks seems impractical.

People say that no observations of quarks or gluons as free particles have been made. When discussing the strong force, people use the term asymptotic behavior in conjunction with force's inferred long-range spatial dependence of $\sim r^0$.

We think such results are not incompatible with notions of $sign'(\Omega)=-1$ for Q-family members. We correlate somewhat free Q-family behavior with conditions that occurred during inflation.

To the extent our work pertains, possibly, attempts to directly observe free quarks would be impractical.

Section 7.3   Black holes and thermal radiation

Abs.7.3.1   We show a means by which black holes can lose energy via thermal radiation.
- This work suggests that thermal radiation by black holes occurs.

In Section 6.5, we discuss examples regarding orbits of photons within black-hole event horizons. Here, we consider an example similar to those examples. Here, for some part of its orbit, the photon's Q-mass exists outside the event horizon. That means that, here, point-1 and point-2 lie outside (or inside but adequately near) the event horizon for the black hole. The cloud of virtual particles associated with the photon can interact with objects outside the event horizon. This next item pertains.

> Tbd.7.3.1   To what extent do photons with orbits that lie partly (but not   (7.3.1)
> wholly) outside a black hole's event horizon provide bases for black-hole
> thermal radiation and/or black-hole dissipation?

Section 7.4   Perihelia shifts and other observations people correlate with general relativity

Abs.7.4.1   We provide quantum-based models correlating with gravitational redshift, gravitational bending of trajectories of light, and shifts of perihelia.

■   This work provides quantum models for known and possibly yet-to-be-observed phenomena that people might try to model via general relativity.

Abs.7.4.2   We discuss the cooling (or redshift) over time of CMB (cosmic microwave background radiation).

■   This work dovetails with our work's not correlating with an expansion of space time.

We continue to develop examples, based on thought experiments Section 6.5 describes.

We consider an example in which the orbiting object is a photon. Here, the hub is a star. Here, point-1 and point-2 lie far from the star. Here, the orbits pass near, but do not touch, the star. Results paralleling those from a previous example [Section 6.5] correlate with blue-shift as the photon moves toward the star and with redshift as the photon moves away from the star. The next items pertain.

- For describing the bending of a path of light by gravity, people can use a model based on Q-mass (instead of a model based on geodesics)   (7.4.1)
- For describing the red-shifting or blue-shifting of light by gravity, people can use a model based on Q-mass (instead of a model based on geodesics)   (7.4.2)

We consider an example in which the orbiting object has non-zero C-mass that is much less than the C-mass for the hub. Here, point-1 and point-2 are suitable for thinking of the object as being a planet orbiting a star (that the hub represents) in a non-circular elliptical orbit. We assume that neither object rotates. We use the symbol L to denote the angular momentum (in units of ℏ) associated with the object's Q-mass. We use the symbol L' to denote an attempt at calculating an angular momentum (in units of ℏ) based on the object's C-mass and motion. At each point in the orbit L>L'. Assume an observer thinks in terms of C-mass. For this observer, an average difference in L−L' (over a complete orbit) correlates with the angular momentum the observer correlates with a perceived perihelion shift. The next item pertains.

Tbd.7.4.1   A model that appropriately considers Q-mass and C-mass dovetails numerically with observed perihelion shifts.   (7.4.3)

We think people can extend such examples to consider other effects - such as some aspects of time dilation - that people model via general relativity. For some of these effects, at least 1 of the interacting objects rotates and the 44G4& analogy to a 22G2& magnetic field pertains. The next items pertain.

Tbd.7.4.2   Determine the extent to which a model that considers Q-mass and C-mass (and possibly forces beyond Standard Model forces and gravity) dovetails with observed phenomena (that people associate with general-relativistic effects) that we do not address in this paper.   (7.4.4)

Tbd.7.4.3   Determine the extent to which a model that considers Q-mass and C-mass (and possibly forces beyond Standard Model forces and gravity) predicts observable yet-to-be-observed phenomena.   (7.4.5)

~ ~ ~

After about 380,000 years ($10^{5.6}$ years [Section 8.1]) after the big bang, a temperature associated with (say, the peak of) the energy-distribution of CMB (an acronym for cosmic microwave background radiation) has decreased. People might say that the universe has expanded and, therefore, the wavelengths of CMB photons have decreased. We propose a different explanation for such cooling of CMB.

As a thought experiment, image an emitter of a CMB photon. We select an emitter that is approximately at rest in a co-moving reference frame. We image that the emitter remains essentially at rest in that reference frame. We imagine an object that - after the photon is emitted - will receive the photon. We imagine that the object will be, at the time of receipt, approximately at rest in a co-moving reference frame.

We imagine that an observer that is at rest in the emitter's reference frame determines a Q-mass for the photon (or, perhaps more correctly, that the observer measures a Q-mass for a similar photon).

At the time of photon emission, the eventual-receiver object lies some non-zero distance (in the emitter's reference frame) from the emitter. In that reference frame, the object is already moving away from the emitter, based on effects of 84G2468&. Various forces, such as 84G2468& lead, in that reference frame, to the object continually accelerating away from the emitter.

When the object absorbs the photon, the object (in effect) detects a photon Q-mass (in the object's reference frame) that is smaller than the Q-mass detected by the original observer.

This effect correlates with CMB cooling. People might use the term Doppler shift to characterize this effect. We note that this Doppler shift differs from local Doppler shifts that people discuss regarding motions of objects within (say) a galaxy or galactic cluster (or, that some people might say, correlate with motions with respect to a cosmic co-moving rest frame).

Section 7.5   The galaxy rotation problem

Abs.7.5.1   A gravitational analog to electromagnetism's magnetic field helps contain material within galaxies.

  ▪   This work points to the possibility that each of 44G4& and dark matter contribute to the forces (beyond those of traditional interpretations of gravity) that keep material within galaxies.

People say that galaxies rotate too rapidly to retain visible material, if the only significant containment force is traditional gravity. People say that dark matter may contribute enough mass and traditional gravitational force to account for the retention of material. People speculate as to the extent other effects (and possibly less dark matter) may help account for the retention of material.

We think that a 44G4& analog to the magnetic field people associate with 22G2& provides such a force. We can think of a (geometric) sphere, centered at the center of a galaxy. The analog field generated by the material inside the sphere affects the (somewhat disk of) material outside the sphere. Each of the inside stuff and the outside stuff rotates in the same direction. The force helps keep stuff outside the sphere from moving into orbits with larger radii. [Section 6.5] (Presumably, effects of the outer stuff on the inner stuff occur and are not as significant as the effects of the inner stuff on the outer stuff.)

The next item suggests possible opportunities for research.

> Tbd.7.5.1   For each of various galaxies, to what extent does cohesion depend   (7.5.1)
> on each of gravity as traditionally interpreted, the gravitational analog to
> electromagnetic magnetic fields, the presence of dark matter, and other
> phenomena?

Section 7.6   The flyby anomaly

Abs.7.6.1   We discuss possible explanations of the flyby anomaly.
  - This work may point to means to test for gravitational effects we predict.

People say that the observed energy of spacecraft that use earth for velocity boosts may differ from anticipated values. Perhaps, to some extent, these observations reflect people's use of C-mass (and not Q-mass) in equations modeling the motions of planets. Perhaps, to some extent, these observations reflect people's use of C-mass (and not Q-mass) in equations modelling a spacecraft's gravitational interactions with non-earth objects before the spacecraft returns to close to earth. Perhaps, to some extent, these observations reflect people's not considering effects of the 44G4& analog to 22G2& magnetic fields.

Section 7.7   Dirac neutrinos and Majorana neutrinos

Abs.7.7.1   Of 3 neutrinos, 2 are Majorana fermions and 1 is a Dirac fermion.
  - This work points to possible needs for new theory regarding neutrino oscillations between Dirac neutrinos and Majorana neutrinos.

Table 4.4.7 indicates that, of 3 neutrinos, 2 are Majorana neutrinos and 1 is a Dirac neutrino. The next table summarizes findings about numbers of zero-charge leptons.

Table 7.7.1   Numbers of zero-charge leptons per peer

Do sibling states correlate with elementary particles?	Number of Dirac zero-charge leptons per peer	Number of Majorana zero-charge leptons per peer	(7.7.1)
no	1	2	(7.7.2)
yes	3	12	(7.7.3)

Section 7.8   Vacuum zero-point energy, mass of the quantum vacuum, and the cosmological constant

Abs.7.8.1   For a photon states, $\xi'=0$.
  - This work provides a way to avoid dealing with the infinity people associate with a sum of photon ground-state energies.

In traditional mathematical physics, the ground state energy (ξ' (1/2)) is both positive and not invariant. Regarding traditional physics, people say that the sum (over photon states) of ground state energies is unbounded.
In our work, Œ = 0 for all representations of all states. The sum of all such Œ is 0.
The next item pertains. (Item (4.3.2) and discussion following that item provide a parallel case.)

Gss.7.8.1 ξ' = 0 correlates with C-mass=0 for G-family bosons.   (7.8.1)

People speculate that zero-point energy could cause an overly large cosmological constant. People wonder why predicted mass of a supposed quantum vacuum has little effect on the expansion of the universe. (People use the term vacuum catastrophe.)

The next items pertain. (Regarding item (7.8.3), our quantum-based work may be a basis for explaining phenomena people correlate with general relativity and may not need the concept of a cosmological constant. Regarding item (7.8.4), we have a non-traditional approach to phenomena people correlate with the term expansion of the universe.)

- People can consider that unoccupied G-family boson states do not contribute to concepts people term vacuum zero-point energy and mass of the quantum vacuum   (7.8.2)
- Based on item (7.8.2), people might say that unoccupied G-family states do not contribute to the cosmological constant   (7.8.3)
- Based on item (7.8.2), people might say that unoccupied G-family states do not contribute to a possible vacuum catastrophe   (7.8.4)

Section 7.9   Nuclear physics and the nature of the nuclear force

Abs.7.9.1   Perhaps people will find work herein useful regarding understanding the nuclear force or nuclear physics.
- This work points to opportunities to enhance theory and models pertaining to the nuclear force and nuclear physics.

The next items pertain. People say that traditional models (such as the shell model) for nuclear physics include concepts based on harmonic-oscillator mathematics.

Tbd.7.9.1   To what extent might people use O-family physics to refine, parallel, or supplant models people use regarding the nuclear force and/or nuclear physics?   (7.9.1)

Tbd.7.9.2   To what extent might people use mathematical models (that this paper includes) to refine, parallel, or supplant models for nuclear physics?   (7.9.2)

Tbd.7.9.3   To what extent might people use mathematical models (that this paper includes) and/or O-family physics to refine, parallel, or supplant models for neutron stars?   (7.9.3)

Part 8   Other topics

Section 8.0   Introduction and summary

Abs.8.0.1   We discuss possible correlations (or lack thereof) between observations and theories, strengths for interactions mediated by G-family members, and the question of extra dimensions.
- This work points to opportunities for observational research.

In Part 8, we discuss some observational and theoretical topics not addressed above. An example of an observational topic is observed ratios of densities for dark energy and non-dark-energy. An example of a theoretical topic is extra dimensions. We point to possibilities for a perturbation theory.

Section 8.1   Why observations do not need to correlate with a 3:1 density ratio of DEF to NDEF

Abs.8.1.1   We provide a possible explanation for the observed density ratio of dark-energy stuff to dark matter plus baryonic matter being less than 3:1.
- This work points to the desirability of finding means (other than via CMB {that is, cosmic microwave background} data) to gain insight about dark-energy stuff.

People estimate ratios of contributions to the density of the universe of baryonic matter, dark matter, and dark energy. People base estimates on data about CMB (that is, cosmic microwave background radiation). Traditional physics considers that observed CMB photons have existed since near the time of the big bang. Physics considers that before $10^{5.6}$ years after the big bang, such photons interacted significantly with ionized plasma. Around $10^{5.6}$ years after the big bang, the plasma ceased to be ionized. Much CMB travels today.

Data suggest late-time (or, secondary) anisotropies. Late-time refers to any time after the above-mentioned plasma ceased to be ionized. Relevant processes for generating non-uniformities in CMB may feature photon scattering by free electrons (Thompson scattering), scattering by clouds of high-energy electrons (Compton scattering by hot electron gases), and frequency-shifting because of changing gravitational fields (integrated Sachs-Wolfe effect). [Ref.8.1.1]

Analyses of CMB data determine, in effect, aspects of clumping. We use the term clumping to denote non-homogeneity in the distribution of energy or matter. Quarks+BMS6(0) clumps correlate with atomic nuclei, atoms, planets, stars, solar systems, clouds of electrons, and so forth. NDEF clumps include galaxies, galactic clusters, and so forth. Possibly, DEF undergoes clumping. [For terminology, see Section 3.4 and Section 5.1]

For MP(1)+MP(2)+MP(3) to have fully impacted CMB, activity the next items list may need to have occurred. We are uncertain as to the extent 0H0 interactions might contribute significantly. To the extent 0H0-mediated interactions provide significant impact regarding these activities, people should add 0H0 to the interactions in items (8.1.2), (8.1.3), and (8.1.4). Also, the extent 0H0 interactions pertain, there might be some contribution to perceived dark energy based on peers MP(4) through MP(7).

- MP(2) and MP(3) have clumped   (8.1.1)

- 8kG…8& interactions have induced clumping in MP(0) and MP(1) based on   (8.1.2)
  clumping in MP(2) and MP(3)
- 6kG…6& and 8kG…8& interactions have induced clumping in MP(0) based   (8.1.3)
  on clumping in MP(1)
- 4kG…4&, 6kG…6&, and 8kG…8& interactions have induced clumping in   (8.1.4)
  quarks+BMS6(0) based on clumping in MP(0) dark matter
- CMB photons (some of the 22G2& associated with MP(0)) have reacted to   (8.1.5)
  the following
  - Clumping of NDEF objects
  - Effects of NDEF clumping on the gravitational field (and other fields)
    germane to CMB photons

Thus, the observed ratio of ~2,2:1 need not be inconsistent with an actual ratio of 3:1.

~ ~ ~

Ref.8.1.1   J. Beringer et al. (Particle Data Group), *Phys. Rev. D86*, 010001 (2012).
             (http://pdg.lbl.gov/2012/reviews/rpp2012-rev-cosmic-microwave-background.pdf)

Section 8.2   G-family interaction strengths

Abs.8.2.1   We sketch relationships between strengths of some G-family forces.
- This work points to observational and theoretic opportunities to better understand G-family forces.
Abs.8.2.2   We explore concepts related to the possibility that perturbation theories pertain.
- This work provides hints that people can use perturbation theory to estimate some quantities.

In conjunction with item (3.5.7), we introduce a concept of channels pertaining to interactions mediated by G-family bosons.

Here, we interpret the notion of channels mathematically for 66G6& and 88G8&. Thereby, we provide another instance of math that correlates with a lack of physics-relevance for G-family states jkG%& for which A`∈%. And, we provide a means for people to think about channels.

The next table shows a concept for something we term a closed oscillator pair.

Table 8.2.1   Definition of closed oscillator pair
- We call the pair of oscillators EjR and EjL (for j being an even positive   (8.2.1)
  integer) closed if and only if people can characterize it as being equivalent
  to a state for which N(EjR) + N(EjL) = −1
  - An example is (| N(EjR)=−1 , N(EjL)=0 > + | N(EjR)=0 , N(EjL)= −1 >) /
    $2^{1/2}$
- We call the pair of oscillators PjL and PjR (for j being an even positive   (8.2.2)
  integer) closed if and only if people can characterize it as being equivalent
  to a state for which N(PjL) + N(PjR) = −1
  - An example is (| N(PjL)=−1 , N(PjR)=0 > + | N(PjL)=0 , N(PjR)= −1 >) /
    $2^{1/2}$

The next items pertain.

- For a closed oscillator pair, Œ=0   (8.2.3)
- We symbolize an oscillator's participation in a closed pair via the symbol *   (8.2.4)

The next table depicts channels. We associate each channel with a pair of oscillator pairs. Each pair of pairs features, for some j = 2, 4, 6, or 8, oscillators EjR, EjL, P(j+2)L, and P(j+2)R.

Table 8.2.2   Ground states and channels for jjGj& solutions (a possible depiction)   (8.2.5) / (8.2.6)

E A`R	E A`L	E 8R	E 8L	E 6R	E 6L	E 4R	E 4L	E 2R	E 2L	P 0	P 0	P 2L	P 2R	P 4L	P 4R	P 6L	P 6R	P 8L	P 8R	P A`L	P A`R	Solution
*	*	*	*	*	*	*	*	0	–1	0	0	*	*	*	*	*	*	*	*			22G2& (8.2.7)
*	*	*	*	*	*	#	#	0	–1	#	#	0	0	*	*	*	*	*	*			44G4& (8.2.8)
*	*	*	*	#	#	#	#	0	–1	#	#	#	#	0	0	*	*	*	*			66G6& (8.2.9)
*	*	#	#	#	#	#	#	0	–1	#	#	#	#	#	#	0	0	*	*			88G8& (8.2.10)
#	#	#	#	#	#	#	#	0	–1	#	#	#	#	#	#	#	#	0	0			A`A`GA`& (8.2.11)

Based on a possible similarity between 22G2& ground states and W-family ground states [Table 6.3.3], the next table provides a possible alternative depiction of channels. (Here, people might consider showing, for each row, N(E2R) = N(E2L) = N(E0) = N(P0) = 0.

Table 8.2.3   Ground states and channels for jjGj& solutions (a possible alternative depiction)   (8.2.12) / (8.2.13)

E A`R	E A`L	E 8R	E 8L	E 6R	E 6L	E 4R	E 4L	E 2R	E 2L	P 0	P 0	P 2L	P 2R	P 4L	P 4R	P 6L	P 6R	P 8L	P 8R	P A`L	P A`R	Solution
*	*	*	*	*	*	*	*			0	–1	0	0	*	*	*	*	*	*	*	*	22G2& (8.2.14)
*	*	*	*	*	*	#	#			0	–1	#	#	0	0	*	*	*	*	*	*	44G4& (8.2.15)
*	*	*	*	#	#	#	#			0	–1	#	#	#	#	0	0	*	*	*	*	66G6& (8.2.16)
*	*	#	#	#	#	#	#			0	–1	#	#	#	#	#	#	0	0	*	*	88G8& (8.2.17)
#	#	#	#	#	#	#	#			0	–1	#	#	#	#	#	#	#	#	0	0	A`A`GA`& (8.2.18)

The next item provides a possible narrative about a role for channels.

Gss.8.2.1   People might think of the excitation of a G-family boson as including (8.2.19) a step in which oscillator P0 is excited –N(P0) times and the QE-like oscillator pair for 1 channel is also excited –N(P0) times. Another step involves the P0 excitements being distributed to Pjk oscillators (with each j≥2 and each k being either R or L) and the QE-like channel excitements being transferred to oscillator E0.

People might say that such a channel-related narrative need not pertain to excitation of W-, H-, and O-family bosons. For those bosons, N(P0) is non-negative.

~ ~ ~

While solutions 66G6& and 88G88& might not correlate with elementary particles, we state the next items as representing possibly useful math.

Gss.8.2.2   The expression $(3/2)(\beta^{(j\times6)})^2$ could represent a ratio of the strengths (8.2.20) of 22G2& to 66G6& for interactions between 2 electrons. Here, j might be 2.

Gss.8.2.3   The expression $(2/1)(\beta^{(k \times 6)})^2$ could represent a ratio of the strengths   (8.2.21)
of 22G2& to 88G8& for interactions between 2 electrons. Here, k might be
3 or 4.

We think the next items pertain. Here, F' denotes the magnitude of the per-channel force for interactions between 2 electrons. The F'(66G6&) and F'(88G8&) correlate with math and might not correlate with nature. Clearly f'(2)=1.

- F'(22G2&) = f'(2) F'(22G2&)   (8.2.22)
- F'(42G24&) = f'(4) F'(22G2&) F'(44G4&)   (8.2.23)
- F'(64G246&) = f'(6) F'(22G2&) F'(44G4&) F'(66G6&)   (8.2.24)
- F'(84G2468&) = f'(8) F'(22G2&) F'(44G4&) F'(66G6&) F'(88G8&)   (8.2.25)

The next item pertains. Perhaps, j = 2. Perhaps, k = 3 or 4.

Tbd.8.2.1   Determine the f'(4), f'(6), j [Item (8.2.20)], k [Item (8.2.21)], and   (8.2.26)
f'(8) that pertain to the magnitudes of G-family #E=0 forces.

~ ~ ~

Work above features interactions between 2 charged generation-1 1L elementary particles (that is, electrons and/or positrons). The next item pertains.

- Vertex strengths scale per particle properties   (8.2.27)

For example, as part of calculating a strength for a vertex that pertains for a jkG%&-mediated interaction, people should multiply the strength of a vertex pertaining to generation-1 charged 1L by the ratio PRjkG%(for the interacting fermion)/PRjkG%(for an electron).

The next item pertains. (For some results regarding attraction or repulsion, see Section 2.1.)

Tbd.8.2.2   Determine strengths and directions (attraction or repulsion) for G-   (8.2.28)
family forces other than 22G2& and 44G4&.

~ ~ ~

We discuss the matter of the 66G6&, 86G6&, and 88G8& solutions. The next items pertain. [Table 3.5.3 and Table 3.5.4]

Gss.8.2.4   PR66G6 correlates with magnetic moment.   (8.2.29)
Gss.8.2.5   PR88G8 correlates with 2S.   (8.2.30)

The next table pertains. People might say that the terms PR66G6 and PR88G8 (respectively) pertain if and only if solutions 66G6& and 88G6& correlate with G-family particles. Here, we propose transformation symmetries. People can use the symmetries when thinking about how quantities differ when seen by different observers. For example, for PR22G2, the traditional charge-and-current 4-vector pertains. We discuss the proposed mass-and-mass-current 4-vector in Section 6.5. The symmetries also pertain to constructing operators for use in current algebras within quantum field theory. The term axial vector provides another term for pseudovector. We do not resolve in the paper whether pseudovector or scalar pertains for 66G6&.

Table 8.2.4   Property, object property, and symmetry for 4 jjGj& solutions

Property	Object property	Solution	Symmetry	
				(8.2.31)
PR22G2	charge	22G2&	4-vector	(8.2.32)
PR44G4	Q-mass	44G4&	4-vector	(8.2.33)
(PR66G6)	magnetic moment	66G6&	pseudovector or scalar	(8.2.34)
(PR88G8)	2S	88G8&	scalar	(8.2.35)

~ ~ ~

The next table pertains. Here, for example, each of 22G2& (for which #E=0) and 44G4& (for which #E=2) correlates with a 4-vector. For example, 22G2& correlates with the charge-and-current 4 vector. Also, each of the other #E=0 G-family forces correlates with a 4-vector. We do not resolve in the paper whether pseudovector or scalar pertains for #E=4.

Table 8.2.5   Transformation symmetries pertaining to G-family mediated interactions

#E	QE-like symmetry	Instances	Transformation symmetry	
				(8.2.36)
0	SU(3)	8	4-vector	(8.2.37)
2	SU(3)	8	4-vector	(8.2.38)
4	SU(5)	24	pseudovector or scalar	(8.2.39)
6	SU(7)	48	scalar	(8.2.40)

~ ~ ~

In this subsection, we explore possibilities that perturbation techniques pertain. The next item sets an overall agenda, which we do not fulfill herein.

> Tbd.8.2.3   Develop a suitable perturbation theory (possibly based on (8.2.41)
> something like Feynman diagrams) for G-family interactions.

People might say that items (3.5.7), (8.2.20), and (8.2.21) correlate with a possible perturbation-theory series. People might say that the parameter in the series is β^{-1}.

We note a possible approximation for α, the fine-structure constant. People might correlate this work with the j2G2& series of G-family bosons. [Item (4.3.1)] We note that, for example, while solution 66G6& might not correlate with an elementary particle, we think that solution 62G2& does correlate with an elementary particle.

The next items define the approximation.

$$\alpha \approx \Sigma \text{ (channel ratio) } \kappa^{\gamma} (2\pi)^{\gamma'} \qquad (8.2.42)$$
$$\text{A channel ratio denotes the ratio of channels for jkG2\& to channels for 22G2\&} \qquad (8.2.43)$$
$$\kappa = 2 \qquad (8.2.44)$$

Here, we choose γ for κ^{γ} to match series that feature powers of β. We choose $\kappa = 2$ to correlate with the opening (that is, unclosing; see Table 8.2.2 and Table 8.2.3) of QP-like oscillator pairs. Here, vis-à-vis items (3.5.7), (8.2.20), and (8.2.21), we show results for k=3. (We do not provide a result item (8.2.48) for k=4.) We choose γ' to obtain an approximate result. The next table shows numbers.

Table 8.2.6   A sum, approximating the fine-structure constant, of terms

channel ratio	γ in κ^{γ}	γ' in $(2\pi)^{\gamma'}$	Single term	Σ = cumulative sum of terms	$(\Sigma - \alpha) / \alpha$	(8.2.45)
3/4	-12	2	7.22871×10^{-3}	7.2287×10^{-3}	-9.4059×10^{-3}	(8.2.46)
2/4	-24	4	4.64483×10^{-5}	7.2752×10^{-3}	-3.0408×10^{-3}	(8.2.47)
1/4	-36	8	8.83688×10^{-6}	7.2840×10^{-3}	-1.8299×10^{-3}	(8.2.48)

The next item pertains. Perhaps, the most relevant interactions to consider include 84G26&, 82G46&, and 82G68&. Each of these jkG%& interactions shares the aspect 6∈% with the 66G6& solution.

Tbd.8.2.4   Assuming that PR66G6 correlates with magnetic moment,   (8.2.49)
             determine the extent to which people can use such a perturbation theory
             [Item (8.2.41)] to estimate magnetic-moment anomalies. [Items (10.5.14)
             and (10.5.15)].

~ ~ ~

People might ask regarding possible significance of β^{-1}. The next items pertain.

- $\beta^{1/3} = e^{e}\times(1-x")$, with $x" \approx 3\times10^{-4}$ (8.2.50)
  - Here, $(\pi/2)/\cos^{-1}(1-x") \approx 65$
- $\beta^{1/3} = e^{e\times(1-x")}$, with $x" \approx 1\times10^{-4}$ (8.2.51)
  - Here, $(\pi/2)/\cos^{-1}(1-x") \approx 107$

The next items pertain. (Compare with items (4.2.14) and (4.2.15).) Regarding item (8.2.54), perhaps $x"\neq0$ correlates with a mixing angle pertaining regarding the 66G6&, 86G6&, and 88G8& solutions and regarding 2 SU(3) symmetries. One such symmetry is the QE-like instance of SU(3) that correlates with the existence of 8 instances of 44G4&. The other such symmetry is the QP-like SU(3) symmetry that correlates with matching 44G4& QP-like oscillators with 3 space-time coordinates.

- $\beta^{1/3}$ is somewhat smaller than e^{e} [Items (8.2.50) or (8.2.51)]   (8.2.52)
Tbd.8.2.5   To what extent does $\beta^{1/3} \approx e^{e}$ correlate with a symmetry (or other   (8.2.53)
             concept) that people would find meaningful?
Tbd.8.2.6   To what extent does $|\beta^{1/3} - e^{e}| \neq 0$ correlate with a mixing angle that (8.2.54)
             correlates with the 66G6&, 86G6&, and 88G8& solutions and 2 instances of
             SU(3) symmetry?

~ ~ ~

We discuss an alternative concept regarding items (8.2.50) and (8.2.51).

Perhaps $x"=0$ correlates with C-mass. If so, the next items might pertain. Item (8.2.55) implies a value for C-mass(electron). Item (8.2.56) implies a value for C-mass(tauon).

$$(4/3)e^{36e} = \{(q_{e})^{2}/(4\pi\varepsilon_0)\} / \{G_{N} \,(\text{C-mass(electron)})^{2}\} \qquad (8.2.55)$$
$$e^{3e} = \text{C-mass(tauon)} / \text{C-mass(electron)} \qquad (8.2.56)$$

Such could lead to neutrino C-masses that are lower than the neutrino Q-masses Table 4.4.7 states. The next table pertains.

Table 8.2.7   Possible C-masses for electrons, tauons, and neutrinos

Particle	Possible C-mass (eV/c^2)	Q-mass (eV/c^2)	(C-mass − Q-mass) / Q-mass	(8.2.57)
electron	5.084×10^5	5.110×10^5	-5.1×10^{-3}	(8.2.58)
tauon	1.769×10^9	1.777×10^9	-4.3×10^{-3}	(8.2.59)
M"=−6 neutrino (Dirac or Majorana)	4.2×10^{-2}	5.8×10^{-2}	-2.7×10^{-1}	(8.2.60)
M"=−3 neutrino (Majorana)	1.25×10^{-1}	1.25×10^{-1}	-6.0×10^{-3}	(8.2.61)

~ ~ ~

We close this section by referring to the unfilled agenda we state in item (8.2.41).

Section 8.3   O-family masses and leptoquark experiments

Abs.8.3.1   We contrast information regarding the O-family and hypothetical leptoquarks.
- Experimental lower bounds on leptoquark masses might be relevant to discussions pertaining to O-family phenomena.

People use the term leptoquark when discussing various concepts for bosons and interactions people do not include in the Standard Model. Perhaps, some of those concepts dovetail with aspects of the O-family and of O-family-related phenomena.

Work above suggests lower bounds for energies needed to produce O-family bosons. [Table 4.2.3]

We note some results regarding leptoquark physics. Experimental results provide lower bounds for masses of various hypothesized leptoquarks. The next item notes a range of reported lower bounds (in GeV/c^2 and with minimum confidence levels of 95%). [Ref.8.3.1]

$$226 - 685 \text{ GeV/c}^2 \qquad (8.3.1)$$

~ ~ ~

Ref.8.3.1   J. Beringer et al. (Particle Data Group), *PR D86*, 010001 (2012) and 2013 partial update for the 2014 edition (URL: http://pdg.lbl.gov).

Section 8.4   Extra dimensions, plus solutions for which D$_P$ need not equal 3

Abs.8.4.1   We explore solutions for which D$_P$ would not be 3.
- This work provides impetus for focusing on D$_P$=3 and for not expecting useful results for other values of D$_P$.

Abs.8.4.2   We explore the topic of the extent to which nature exhibits more than 4 dimensions.
- This work provides the possibility that use of 4 space-time coordinates suffices for models describing much physics.

Much of the work above correlates with $\Omega = \pm S(S+D_P-1)$, for $D_P=3$ and for some S in which 2S is a non-negative integer. Here, we explore possibilities for obtaining similar values of Ω via the use of values other than 3 for D_P.

The next item shows a formula pertaining to spin values that could arise for a choice of $D_{*P}=1$.

- $\Omega = \pm S(S+D_{*P}-2) = \pm S(S-1)$, with ...   (8.4.1)
  - $D_{*P}=1$
  - $S(S-1) = 0, 0, 2, 6, ...$ respectively for $S = 0, 1, 2, 3, ...$
  - $S(S-1)$ would be $-1/4$ for $S = 1/2$
  - $S(S-1) = 3/4, 15/4 ...$ respectively for $S = 3/2, 5/2, ...$

The next items discuss solutions for $\nu < 0$.

- $\nu = -1/2$ correlates with edge solutions   (8.4.2)
- No $\nu < 0$ inside solutions exist   (8.4.3)

We interpret item (8.4.3) as correlating with non-existence of a field for otherwise potentially possible fermion particles corresponding to item (8.4.2). Also, we interpret results as correlating with non-existence of $D_{*P}=1$ bosons.

For odd $D_{*P} \geq 5$, Ω cannot match $\pm S(S+1)$ for which 2S is a non-negative integer.

All as-yet-known physics exhibits spins for which $S(S+1)$ pertains with 2S being a non-negative integer. Apparently, $D_{*P}=3=D_P$ correlates with all known observations.

$$\sim \sim \sim$$

Work above models physics based on 1 QE-like dimension and 3 QP-like dimensions. People might note that we use solutions (to QP-like equations) for which for which D>3. We note that these solutions correlate with $D_P=3$ and with use of space-time coordinates.

The next item pertains.

- To the extent models we show provide a basis for all natural phenomena,   (8.4.4)
  people do not need to use more than 4 space-time coordinates

Section 8.5   Our work, special relativity, and the Standard Model

Abs.8.5.1   We discuss overlaps and differences in phenomena modeled by work in this paper, special relativity, and the Standard Model.
- Work in this paper is adequately compatible with special relativity.
- People may be able to extend work in this paper to include parts of the Standard Model that work in this paper does not address.

In this subsection, we discuss compatibility between special relativity and our quantum-based approach.

People might find compatibility between special relativity and work in this paper regarding QE-like phenomena (or, $\Omega > 0$ phenomena). For example, for $\Omega > 0$, we derive, for each elementary particle, $E^2 - c^2 P^2$ = constant ≥ 0 as an expression involving quantum operators.

People might say that, generally, $\Omega < 0$ phenomena pertain significantly on small QE-like scales and on small QP-like scales. Here, small correlates with the QP-like size (or, diameter) of a nucleon. People might say that an exception is the lifespan of a quark. But, the existences of free (or superconducting-like) 1Q particles and free (or, superconducting-like) 3Q particles would be limited to conditions such as those that existed before the formation of individual nucleons.

The next item pertains.

> • Perhaps people can be content with the notions that ...   (8.5.1)
>   • Special relativity pertains to classical physics
>   • Traditional quantum mechanics dovetails with that aspect of classical physics
>   • Our quantum approach dovetails with that aspect of classical physics

~ ~ ~

In this subsection, we correlate our representations for Standard Model elementary particles with an SU(3)×SU(2)×U(1) symmetry.

People say that the Standard Model correlates with an SU(3)×SU(2)×U(1) symmetry. Some Standard Model work features considering currents and their behavior via use of space-time coordinates.

Our work uses space-time coordinates. People might say we have yet to match fully those coordinates with similar coordinates that people would correlate with the Standard Model. For example, for #E=0 G-family particles, we make special use of P0. Some people might correlate that special use with a possible need to align x0 with p0 with P0. [Table 1.1.6]

The next item pertains.

> • Vis-à-vis Standard Model use of space-time coordinates and our use of   (8.5.2)
>   space-time coordinates, a QP-like SU(3) symmetry pertains

The next items pertain.

> • Our representation for W-family bosons correlates with a QP-like   (8.5.3)
>   SU(2)×U(1) symmetry
> • Our representation for 2Y bosons correlates with a QE-like SU(2)×U(1)   (8.5.4)
>   symmetry
> • Our representation for 22G2& correlates with a QP-like SU(2)×U(1)   (8.5.5)
>   symmetry

The next item pertains. Here, we combine item (8.5.2) with each of items (8.5.3), (8.5.4), and (8.5.5).

> • For all Standard Model forces except 0H0, our work correlates with an   (8.5.6)
>   SU(3)×SU(2)×U(1) symmetry

People might say that our representation for 1L Dirac-like particles correlates with item (8.5.3) and with an SU(3)×SU(2)×U(1) symmetry.

People might say that our representation for 1Q particles correlates with item (8.5.4) and with an SU(3)×SU(2)×U(1) symmetry.

The next item pertains.

> Gss.8.5.1   People would say that statements above in this section correlate   (8.5.7)
>   with our work's correlating with the SU(3)×SU(2)×U(1) symmetry people
>   associate with the Standard Model.

~ ~ ~

Item (8.5.7) notes relationships between our representations, Standard Model elementary particles, and the SU(3)×SU(2)×U(1) symmetry people correlate with the Standard Model.

The next items pertain.

- People might say that the Standard Model is not incompatible with work in this paper   (8.5.8)
  - Work in this paper correlates with phenomena beyond those with which the Standard Model deals
    - An example of such phenomena is some behavior of non-zero-mass $\Omega<0$ elementary bosons
  - Work in this paper correlates with all Standard Model elementary particles and with the Standard Model symmetry SU(3)×SU(2)×U(1)
- People might say that the Standard Model and special relativity are compatible   (8.5.9)
- People might say that this paper does not address some phenomena people address via the Standard Model   (8.5.10)
  - An example of phenomena this paper does not adequately address is compound particles (such as protons)

Gss.8.5.2   People can extend work in this paper to include (or merge with) the Standard Model.   (8.5.11)

Section 8.6   Other topics

Abs.8.6.1   We note some topics we do not address.

- These unaddressed topics may represent opportunities for further research.

The next items pertain.

Table 8.6.1   Other possible opportunities for research

Tbd.8.6.1   Provide theory correlating with tauon and muon decay rates.   (8.6.1)

Tbd.8.6.2   Pinpoint important interaction vertices pertaining to neutrino oscillations.   (8.6.2)

Tbd.8.6.3   Try to address the topic of arrow of time.   (8.6.3)
- For example, explore the extent to which the following correlate
  - Arrow-of-time concepts
  - Effects of the force 84G2468&
  - Lack of possibilities for exciting the state 0H0' (the $\Omega<0$ counterpart to 0H0, which correlates with the Higgs boson)
  - Entropy

Tbd.8.6.4   To what extent does the existence of the 0H0 particle correlate with the existence of elementary particles with non-zero mass?   (8.6.4)

Tbd.8.6.5   To the extent work in this paper provides a complete list of elementary particles, to what extent might people say that nature does not exhibit magnetic monopoles?   (8.6.5)

Tbd.8.6.6   Look for, based on work above, possibly useful physics for bases to   (8.6.6)
develop new observational approaches to infer or experiments to detect
gravitons or gravitational waves.

Tbd.8.6.7   To what extent should people correlate the usefulness of space-time   (8.6.7)
coordinates with usefulness for concepts of space time? To what extent
can people describe nature (or, at least, quantum phenomena) without
assuming space time has properties? To what extent can people describe
nature (or, at least, quantum phenomena) without assuming space time
exists?

Part 9   Conclusions and extensions

Section 9.0   Introduction and summary

Abs.9.0.1   We provide possible updates to traditional narratives about topics in mathematics,
mathematical physics, and physics.
- This work points to bases for understanding, discussing, and extending work in this paper.

In Part 9, we list changes (that people might say our work suggests) to traditional narratives. The narratives involve areas of mathematics, mathematical physics, elementary-particle physics, astrophysics, and the evolution of the universe (cosmology). For each area, we also note possible opportunities for research.

The next item pertains.

> Tbd.9.0.1   Develop simpler bases for and/or descriptions of work this paper   (9.0.1)
> describes.

Section 9.1   Changes to narratives about mathematics

Abs.9.1.1   We discuss possible updates to traditional narratives regarding mathematics.
- The changes suggest possible opportunities for research.

The next table lists possible changes to traditional narratives regarding mathematics.

Table 9.1.1   Possible changes to traditional narratives regarding mathematics
- Mathematics for multidimensional isotropic quantum harmonic oscillator   (9.1.1)
equations includes solutions for which people might say that single-
dimension component oscillators exhibit negative quantum numbers

The next table lists possible opportunities for mathematics research.

Table 9.1.2   Possible opportunities for research regarding mathematics
> Tbd.9.1.1   Develop practical applications of harmonic-oscillator math beyond   (9.1.2)
> traditional applications and applications we discuss.
> Tbd.9.1.2   Explore a broader range of solutions for a broader interpretation of   (9.1.3)
> the term multi-dimensional harmonic-oscillator math.
> - Possibly, 1 or more of the following pertain
>   - Broader ranges of values (possibly including complex numbers) for
>   parameters such as D_P, Ω, v, S, and D
>   - Perhaps, people can use just the algebraic solutions (and not use wave-
>   function representations)
>   - Each of the sets of QE-like and QP-like oscillators has an even-integer
>   number of members
>   - More than just the 2 sets of oscillator indices for which (sets) we use the
>   terms QE-like and QP-like
>   - Less restrictive concepts than those we correlate with isotropic

Section 9.2   Changes to narratives about mathematical physics

Abs.9.2.1   We discuss possible updates to traditional narratives regarding mathematical physics.
  ▪   The changes suggest possible opportunities for research.

The next table lists possible changes to traditional narratives regarding mathematical physics.

Table 9.2.1   Possible changes to traditional narratives regarding mathematical physics
  • Harmonic-oscillator math provides an alternative basis (to starting from   (9.2.1)
    action and Lagrangian math) for modeling interactions between elementary
    particles (and, more generally, between objects) and for modeling (at least)
    some aspects of motion
  • People may want to take care about applying classical-physics thinking (for   (9.2.2)
    example, some uses of the concept of limits based on the speed of light) to
    $\Omega < 0$ quantum phenomena
  • Mathematics that is adequately quantum-oriented provides means to model   (9.2.3)
    all-known elementary-particle, astrophysics, and cosmology phenomena …
      • Using 4 space-time coordinates in ways such that people would say that
        (Minkowski-like) flatness pertains
      • Using a gravitational analog to the electromagnetic charge-and-current 4-
        vector
      • Using a gravitational analog to the electromagnetic magnetic field (as
        well as a gravitational analog to the electromagnetic electric field)
      • Using concepts of force-governed motion (and not explicitly using
        concepts of geodesics)
      • Treating all objects similarly
          • For example, when considering trajectories, not ignoring the energy
            correlating with a photon or not ignoring the mass of a planet
  • Quantum-based math provides useful ways to describe concepts such as   (9.2.4)
    spin and such as helicity, chirality, and/or handedness

The next table lists possible opportunities for mathematical-physics research.

Table 9.2.2   Possible opportunities for research regarding mathematical physics
  Tbd.9.2.1   To what extent can people use quantum-oscillator-based math to   (9.2.5)
              describe phenomena for which people use classical physics,
              electrodynamics, traditional quantum mechanics, the Standard Model,
              quantum chromodynamics, and/or general relativity?
  Tbd.9.2.2   To what extent can people use results based on quantum-oscillator-   (9.2.6)
              based math to substitute quantum-based approximations for classical-
              physics-based approximations in simulations (for example, simulations of
              interactions between molecules; or, other simulations for which people
              might say that efforts span classical-physics/quantum-physics
              boundaries)?
  Tbd.9.2.3   To what extent can people use quantum-oscillator-based math   (9.2.7)
              reproduce Standard Model math?
  Tbd.9.2.4   To what extent, if any, does a quantum-oscillator-based approach   (9.2.8)
              require use of concepts people might term properties of space time?

Tbd.9.2.5   For phenomena (such as perihelia shifts) we describe qualitatively,   (9.2.9)
develop quantitative treatments that match or predict observations.

Section 9.3   Changes to narratives about elementary-particle physics

Abs.9.3.1   We discuss possible updates to traditional narratives regarding elementary-particle physics.
▪   The changes suggest possible opportunities for research.

The next table lists possible changes to traditional narratives regarding elementary-particle physics.

Table 9.3.1   Possible changes to traditional narratives regarding elementary-particle physics
- We provide a method for cataloguing known elementary particles and for   (9.3.1)
  predicting yet-to-be-discovered elementary particles
  - People might draw parallels between this method (for elementary
    particles) and the periodic table (for atoms).
- Elementary particles can be classified into 7 families   (9.3.2)
  - Currently, the Standard Model does not include members of 1 of the
    families (the O-family)
  - The Standard Model does not include some members of some of the other
    families (the G-, Q-, and Y-families)
  - The Standard Model does not include possible siblings of 1 of the families
    (the L-family)
- Quantum gravity has many similarities to quantum electromagnetism   (9.3.3)
- Photons and gravitons are 2 members of set of zero-mass bosons (that does
  not include gluons) - the G-family
- Theory predicts yet-to-be discovered elementary particles   (9.3.4)
  - Spin-1 bosons include some O-family members and some G-family
    members
  - Spin-2 bosons include the graviton, other G-family members, some O-
    family members, and some Y-family members
  - Spin-3/2 fermions include some Q-family members
- In many circumstances, O-family bosons, like Q-family fermions, cannot   (9.3.5)
  exist as free particles
- An algebraic relationship links the tauon mass, electron mass, charge of an   (9.3.6)
  electron, Coulomb constant, and gravitational constant
- Dark matter includes at least 1 of ...   (9.3.7)
  - Compound particles that include spin-3/2 Q-family fermions and spin-2
    Y-family elementary bosons
  - Siblings of baryonic-matter leptons
- States for higher-mass siblings of leptons may exist   (9.3.8)
  - Possibly, such states depleted early in the history of the universe
- Dark-energy stuff consists of peers of baryonic matter + dark matter   (9.3.9)

- People can associate the concepts of C-mass and Q-mass with objects   (9.3.10)
  - For example, a photon has 0 C-mass and has Q-mass that correlates with the energy (or momentum) that an observer would attribute to the photon
  - Q-mass correlates with the existence of virtual quantum states (for example, fermion-and-anti-fermion pairs) that correlate with an elementary particle or other object
- Of 3 neutrinos, 2 are Majorana fermions and 1 is a Dirac fermion   (9.3.11)
  - This paper suggests possible masses (or mass-math eigenvalues) for neutrinos
- For O-family bosons, this paper computes charges, masses, and lower bounds for threshold energies   (9.3.12)
  - Likely, the threshold energies exceed the lower bounds

The next table lists possible opportunities for elementary-particle physics research.

Table 9.3.2   Possible opportunities for research regarding elementary-particle physics
   Tbd.9.3.1   Detect or rule out (to some confidence level) various elementary   (9.3.13)
                particles to which Table 9.3.1 alludes.
   Tbd.9.3.2   Estimate and/or measure quantities related to phenomena   (9.3.14)
                involving spin-3/2 fermions.
   Tbd.9.3.3   Measure more accurately the tauon mass (m_{tauon}) and the   (9.3.15)
                gravitational constant (G_N); thereby, help clarify validity of theory we
                propose.
   Tbd.9.3.4   Determine the extent to which distinguishing, for gravitons,   (9.3.16)
                between Q-mass-related phenomena and C-mass-related phenomena can
                lead to new tests for the existence and properties of gravitons.
   - Compare with Section 6.4 regarding photons

Section 9.4   Changes to narratives about astrophysics

Abs.9.4.1   We discuss possible updates to traditional narratives regarding astrophysics.
   - The changes suggest possible opportunities for research.

The next table lists possible changes to traditional narratives regarding astrophysics.

Table 9.4.1   Possible changes to traditional narratives regarding astrophysics
   - Quantum-based relativity ...   (9.4.1)
     - Provides models for perihelia shifts and other orbital mechanics people traditionally model via general relativity
     - Provides a force that may have significance regarding the cohesion of galaxies
     - Describes mechanisms for black-hole thermal radiation
     - Includes forces that can lead to black holes producing quasars

The next table lists possible opportunities for astrophysics research.

Table 9.4.2   Possible opportunities for research regarding astrophysics
        Tbd.9.4.1   Verify, detect or rule out (to some confidence level) aspects to which (9.4.2)
            Table 9.4.1 alludes.

Section 9.5   Changes to narratives about cosmology and the evolution of the universe

Abs.9.5.1   We discuss possible updates to traditional narratives regarding cosmology and the evolution of the universe.
   ▪   The changes dovetail with traditional timelines.

The next table lists possible changes to traditional narratives regarding cosmology and the evolution of the universe.

Table 9.5.1   Possible changes to traditional narratives regarding the evolution of the universe
   • Forces 84G2468&, 64G246&, and 42G24& drive large-scale expansion of the (9.5.1) universe
      • Force 84G2468& dominates from before inflation until the end of era FE1
         • The end of era FE1 is later for larger objects
         • For large, currently observable objects, a transition from era FE1 to era FE2 correlates with a transition from era ZE1 to era ZE2 at around 2 billion to 3 billion years after the big bang
      • People need not attribute expansion of the universe to an expansion of space time
   • Inflation correlates with quantum kinematics that people might (9.5.2) characterize as free-motion QP-like phenomena (or, superconductor-like QP-like phenomena) of 3Q fermions
      • Effects of 4Y and 4O bosons pertain
      • Perhaps, inflation ends when the density of 3Q fermions is sufficiently small that 4O particles cannot significantly mediate interactions between 3Q fermions
   • During a time period (that might start before or after inflation) that ends (9.5.3) around the time quarks (1Q particles) form nucleons, ...
      • Lasing of O-family bosons may play a key role in establishing long-term matter/antimatter imbalance
      • The charged lepton states that correlate with BMS6(j) for j>0 depopulate
   • Perhaps the formation of nucleons correlations with the density of 1Q (9.5.4) fermions being sufficiently small that 2O particles cannot significantly mediate interactions between 1Q fermions

The next table lists possible opportunities for cosmology research.

Table 9.5.2   Possible opportunities for research regarding cosmology
        Tbd.9.5.1   Verify, detect or rule out (to some confidence level) aspects to which (9.5.5)
            Table 9.5.1 alludes.

Part 10 Compendia

Section 10.1 Key points and context

Here, we collect a list of key points. Here, we also collect context for the points. This section reports abstracts of sections. (Sections in Part 0 and in Part 10 do not have formal section abstracts.) The next items describe structure we use.

- The label Abs.j.k.n provides a part number j and a section number j.k (10.1.1)
  - For each section j.k for which we show abstract items, we show an item numbered Abs.j.k.1
  - For some sections j.k, we show items numbered Abs.j.k.n, with n≥2
- For each Abs.j.k.n item, we provide ... (10.1.2)
  - Text that directly follows the Abs.j.k.n number
  - 1 or more items, each preceded by a ▪
- The Abs.j.k.n text summarizes the section or provides key points from the section (10.1.3)
- For each item preceded by a ▪, ... (10.1.4)
  - We provide perspective related to the corresponding Abs.j.k.n item
  - People might read the ▪ item as "people might say that ... {text that follows the ▪}"

Part 0	**Introduction and summary**
Part 1	**Elementary particles (not including dark matter or dark energy)**
Abs.1.0.1	We list 7 families of elementary particles.
▪	We report results from using a math model to catalog elementary particles.
Abs.1.1.1	We use 4 harmonic oscillators to describe some aspects of photons.
▪	This work provides a way to discuss observer-invariant aspects of photons.
Abs.1.2.1	We use 8 harmonic oscillators to describe aspects of gravitons.
▪	This work provides a quantum description for gravity.
Abs.1.3.1	The G-family includes all zero-mass elementary particles except Y-family particles.
▪	This work unifies quantum electromagnetism and quantum gravity.
Abs.1.4.1	We summarize math solutions that we correlate with non-zero-mass elementary particles.
▪	This work provides insight into concepts of particles and fields.
Abs.1.5.1	We list all known and some possible non-zero-mass elementary bosons.
▪	This work points to opportunities to discover or infer members of a family of non-zero-mass elementary bosons for which no particles have as yet been identified.
Abs.1.6.1	We list all known non-zero-mass elementary fermions.
▪	This work points to ways to represent quark interaction-vertices and lepton interaction-vertices via harmonic oscillator math.
Abs.1.7.1	The Y-family provides a basis for gluons.
▪	This work provides insight about gluons and related symmetries.
Part 2	**Models that correlate with phenomena**
Abs.2.0.1	We correlate observed phenomena with above-mentioned elementary particles.
▪	This work provides opportunities to resolve various physics-theory problems people associate with the term unsolved.
Abs.2.1.1	The G-family particles 84G2468&, 64G246&, and 42G24& provide for changes in the rate of expansion of the universe.

- This work suggests that forces (and not just the existence and some effects of dark-energy stuff) regulate the observed expansion of the universe.

Abs.2.2.1   Dominance of the G-family forces 84G2468&, 64G246&, and 42G24& correlates with the universe's having little or no large-scale curvature.

- This work suggests that forces and symmetries (and not a critical density) correlate with near or actual flatness of the universe.

Abs.2.3.1   Spin-1 O-family bosons could have catalyzed the current matter/antimatter imbalance.

- This work provides a possible basis for resolving the problem of baryon asymmetry.

Abs.2.4.1   O-family bosons correlate with CPT-related symmetry violations.

- This work points to how to resolve problems related to the sizes of CPT-related symmetry violations exceeding symmetry-violation sizes people correlate with Standard Model physics.

Abs.2.5.1   At least 1 of the forces 42G24& and 84G2468& produces quasars.

- The work provides possible mechanisms that lead to formation of quasars.

Abs.2.6.1   The mass of a tauon may be $1.776814(\sim48)\times10^3$ MeV/c^2.

- This work provides impetus to improve the accuracy of measurements of the mass of tauons and the accuracy of measurements of the gravitational constant.

Abs.2.7.1   Each elementary fermion is part of a 3-generation (or 3-flavor) trio of elementary fermions.

- This work solves a problem regarding providing a math model that correlates with fermions having 3 generations.

Part 3   Mathematics

Abs.3.0.1   We discuss mathematics underlying much of this paper.

- This work opens (and provides applications of) a branch of math within isotropic quantum harmonic oscillators.

Abs.3.1.1   Physics observations include quantized results with which action/Lagrangian math models do not correlate.

- This work provides impetus to develop (from quantum-math bases that do not start from classical-physics action/Lagrangian math) inherently quantum approaches for modeling physics phenomena.

Abs.3.2.1   We discuss mathematics for isotropic quantum harmonic oscillators.

- This work shows new solutions within math related to isotropic quantum harmonic oscillators.

Abs.3.3.1   We note relationships between numbers of generators for groups SU(j) for various j.

- This work may show underutilized arithmetic relationships between numbers of generators for various SU(j) groups.

Abs.3.4.1   We show solutions that might correlate with elementary particles.

- This work provides insight about the use of 3 spatial coordinates for modelling physics.

- This work provides insight regarding relationships between mathematical models for particles and mathematical models for fields.

- This work provides insight regarding which bosons interact with which fermions.

Abs.3.5.1   The ratio of strengths of electromagnetism and gravity correlates with the ratio of masses of the tauon and the electron.

- This work reduces the number of potentially independent fundamental physics constants.

Abs.3.5.2   We provide candidates for phenomena to which jkG...6...& forces couple and to which jkG...8& forces couple.

- This work points to opportunities determine the next (and also, the last) 2 elements in the series charge, mass, ...

Abs.3.6.1   We summarize possible elementary particles that correlate with solutions.

- This work provides themes for experimental research.

Abs.3.7.1   We discuss quantum operators related to motion of elementary particles.

- This work provides a basis for extending our interaction-centric approach (to modeling particle physics) toward a traditional approach to modeling particle kinematics.

Abs.3.7.2   We find quantum numbers related to mass.

- This work provides new insight regarding masses of elementary particles.

Abs.3.7.3   For non-zero-mass elementary particles, $E^2 - c^2P^2 = \text{sign}'(\Omega)\,|m^2|\,c^4$.

- This work correlates with the possibility that the quantum mechanics of quarks or spin-3/2 fermions correlates with inflation.

Abs.3.8.1   We show a way to derive an uncertainty principle for non-zero-mass elementary particles.

- Math we use provides a way to study uncertainty related to base states.

Part 4   Invariant and relative properties

Abs.4.0.1   We predict masses for some elementary particles.
- This work provides opportunities to detect or rule out some possible elementary particles.

Abs.4.1.1   We define C-mass (roughly, classical-physics mass) and Q-mass (roughly, quantum-physics mass).
- This work provides a step toward math for modeling general-relativistic kinematics in a way that accounts for zero-mass particles having energy.

Abs.4.2.1   Masses for O-family bosons are approximately 80.4, 91.2, 105.3, and 113.7 GeV/c^2. Threshold energies for creating minimum numbers of O-family bosons may be at least ≥241, ≥274, ≥402, ≥526, and ≥569 GeV.
- This work points to possible opportunities for experimental research.

Abs.4.2.2   The 202 boson has charge −(1/3)|q$_e$| and the 201 boson has charge +(1/3)|q$_e$|.
- This work provides some understanding of quantum aspects related to fractional charge.

Abs.4.3.1   An arithmetic combination of ℏ, c, and G$_N$ approximates (within less than 8 parts in 100) the mass of a Higgs boson.
- This work possibly relates the masses of W-, H- and O-family bosons to ℏ, c, and G$_N$.

Abs.4.4.1   A formula approximates masses for 6 quarks and 3 charged leptons.
- This work may help substantiate the use of math models that people might say include a negative range for a radial coordinate.

Abs.4.4.2   If sibling states exist, the formula approximates masses for siblings of 3 charged leptons.
- This work may help people direct searches for sibling states.

Abs.4.4.3   Possibly, masses or mass-math eigenvalues (in units of eV/c^2) and types for the baryonic-matter neutrinos are ~0.125 Majorana, ~0.058 Majorana, and ~0.058 Dirac.
- This work may point to numbers correlating with masses or mass-math eigenvalues for neutrinos.

Abs.4.5.1   For G- and Y-family elementary particles, C-mass = 0.
- This work bridges from our work to traditional concepts that photons and gluons have no mass.

Part 5   Dark energy, dark matter, and sibling-state fermions

Abs.5.0.1   We describe dark-energy stuff, dark matter, and possibilities for detectable sibling states.
- This work provides opportunities to resolve various physics-theory problems people associate with the term unsolved.

Abs.5.1.1   Dark-energy stuff consists of as many as 7 peers of the stuff people associate with the combination of baryonic matter and dark matter.
- This work provides a candidate explanation for dark-energy stuff.

Abs.5.2.1   We discuss the possible existence of siblings of leptons.
- Experimental opportunities may exist to verify or rule out sibling states.

Abs.5.3.1   Dark matter consists of at least 1 of spin-3/2 fermions and siblings of baryonic-matter leptons.
- This work points to possible opportunities to observe or infer properties of dark matter.

Part 6   New models for phenomena for which people say traditional theories correlate

Abs.6.0.1   We explore relationships between models in this work and traditional models correlating with spin; helicity, chirality, and handedness; conservation laws; some aspects of photonics; and effects people model via general relativity.
- This work provides a quantum basis for explaining phenomena people correlate with general relativity.
- This work provides a basis for avoiding concepts of curved space times.

Abs.6.1.1   We offer seemingly a straightforward way to model spin.
- We offer math correlating with spin that is more invariant or compelling than some math in traditional approaches.

Abs.6.2.1   We offer seemingly straightforward ways to model helicity, chirality, and/or handedness.
- We offer math correlating with helicity, chirality, and/or handedness that is more invariant or compelling than some math in traditional approaches.

Abs.6.3.1   We show examples of conservation of energy, momentum, and angular momentum.
- These examples show the possibility for deriving the three conservation laws for all interactions.

Abs.6.4.1   We explore roles of relative phenomena and invariant phenomena for photons.
- Parallels, for gravitons, for this work might lead to new ways to test for gravitons.

Abs.6.5.1   We suggest quantum bases for phenomena people might correlate with general relativity.

- This work provides opportunities to determine the extent to which quantum mechanics provides models that underlie aspects of general relativity.

Part 7   Models that may correlate with phenomena

Abs.7.0.1   We provide correlations between this work and various phenomena.
- This work solves or points to how to solve various problems involving the extent to which traditional theories correlate with observed phenomena.
- This work points to opportunities or observational and theoretical research.

Abs.7.1.1   Observations that people term inflation correlate with kinematics of Q-family fermions.
- This work suggests an era in which quarks or spin-3/2 quark-like particles ranged freely or ranged within a fermion sea involving many such particles.

Abs.7.2.1   We discuss lack of observation of free quarks and gluons and we discuss the notion of asymptotic freedom.
- This work supports the notion that observing free quarks seems impractical.

Abs.7.3.1   We show a means by which black holes can lose energy via thermal radiation.
- This work suggests that thermal radiation by black holes occurs.

Abs.7.4.1   We provide quantum-based models correlating with gravitational redshift, gravitational bending of trajectories of light, and shifts of perihelia.
- This work provides quantum models for known and possibly yet-to-be-observed phenomena that people might try to model via general relativity.

Abs.7.4.2   We discuss the cooling (or redshift) over time of CMB (cosmic microwave background radiation).
- This work dovetails with our work's not correlating with an expansion of space time.

Abs.7.5.1   A gravitational analog to electromagnetism's magnetic field helps contain material within galaxies.
- This work points to the possibility that each of 44G4& and dark matter contribute to the forces (beyond those of traditional interpretations of gravity) that keep material within galaxies.

Abs.7.6.1   We discuss possible explanations of the flyby anomaly.
- This work may point to means to test for gravitational effects we predict.

Abs.7.7.1   Of 3 neutrinos, 2 are Majorana fermions and 1 is a Dirac fermion.
- This work points to possible needs for new theory regarding neutrino oscillations between Dirac neutrinos and Majorana neutrinos.

Abs.7.8.1   For a photon states, $\xi'=0$.
- This work provides a way to avoid dealing with the infinity people associate with a sum of photon ground-state energies.

Abs.7.9.1   Perhaps people will find work herein useful regarding understanding the nuclear force or nuclear physics.
- This work points to opportunities to enhance theory and models pertaining to the nuclear force and nuclear physics.

Part 8   Other topics

Abs.8.0.1   We discuss possible correlations (or lack thereof) between observations and theories, strengths for interactions mediated by G-family members, and the question of extra dimensions.
- This work points to opportunities for observational research.

Abs.8.1.1   We provide a possible explanation for the observed density ratio of dark-energy stuff to dark matter plus baryonic matter being less than 3:1.
- This work points to the desirability of finding means (other than via CMB {that is, cosmic microwave background} data) to gain insight about dark-energy stuff.

Abs.8.2.1   We sketch relationships between strengths of some G-family forces.
- This work points to observational and theoretic opportunities to better understand G-family forces.

Abs.8.2.2   We explore concepts related to the possibility that perturbation theories pertain.
- This work provides hints that people can use perturbation theory to estimate some quantities.

Abs.8.3.1   We contrast information regarding the O-family and hypothetical leptoquarks.
- Experimental lower bounds on leptoquark masses might be relevant to discussions pertaining to O-family phenomena.

Abs.8.4.1   We explore solutions for which D_P would not be 3.
- This work provides impetus for focusing on $D_P=3$ and for not expecting useful results for other values of D_P.

Abs.8.4.2	We explore the topic of the extent to which nature exhibits more than 4 dimensions.	
▪	This work provides the possibility that use of 4 space-time coordinates suffices for models describing much physics.	
Abs.8.5.1	We discuss overlaps and differences in phenomena modeled by work in this paper, special relativity, and the Standard Model.	
▪	Work in this paper is adequately compatible with special relativity.	
▪	People may be able to extend work in this paper to include parts of the Standard Model that work in this paper does not address.	
Abs.8.6.1	We note some topics we do not address.	
▪	These unaddressed topics may represent opportunities for further research.	

Part 9   Conclusions and extensions

Abs.9.0.1	We provide possible updates to traditional narratives about topics in mathematics, mathematical physics, and physics.
▪	This work points to bases for understanding, discussing, and extending work in this paper.
Abs.9.1.1	We discuss possible updates to traditional narratives regarding mathematics.
▪	The changes suggest possible opportunities for research.
Abs.9.2.1	We discuss possible updates to traditional narratives regarding mathematical physics.
▪	The changes suggest possible opportunities for research.
Abs.9.3.1	We discuss possible updates to traditional narratives regarding elementary-particle physics.
▪	The changes suggest possible opportunities for research.
Abs.9.4.1	We discuss possible updates to traditional narratives regarding astrophysics.
▪	The changes suggest possible opportunities for research.
Abs.9.5.1	We discuss possible updates to traditional narratives regarding cosmology and the evolution of the universe.
▪	The changes dovetail with traditional timelines.

Part 10   Compendia

Section 10.2   Guesses

Here, we collect a list of guesses.

Part 0   Introduction and summary

Part 1   Elementary particles (not including dark matter or dark energy)

Gss.1.2.1	Oscillators P2L and P2R correlate with charge and with spin-1.
Gss.1.2.2	Oscillators P4L and P4R correlate with mass, with spin-2, and with 2 polarizations for gravitons.
Gss.1.2.3	For gravitons, N(P2L) = N(P2R) = #.
Gss.1.2.4	For the purposes of this work, the 2 P4j oscillators need not correlate (ultimately) with additional QP-like space-time coordinates.
Gss.1.2.5	Oscillators E2R and E2L pertain.
Gss.1.2.6	N(Ej) = 0 (for j = 2R, 2L, and 0) correlates with graviton ground states.
Gss.1.2.7	Œ= 0 for gravitons.
Gss.1.6.1	For elementary fermions, 2S=#P−1.
Gss.1.7.1	One trio of Y-family bosons provides for gluons pertaining to quarks people consider to be matter. The other trio pertains to quarks people consider to be antimatter.

Part 2   Models that correlate with phenomena

Gss.2.1.1	The spatial dependence of the force associated with a G-family boson approximates $r^{2 \times N(P0)}$. Here, r denotes the distance between the centers of property (such as charge {for 22G2&, the electromagnetic force} or mass or mass-energy {for 44G4&, gravity}). Here, N(P0) pertains to the G-family boson.
Gss.2.1.2	For observed astrophysical objects of above some size, era FEk correlates with era ZEn, for 1≤k=n≤3.

Gss.2.1.3   84G2468& repels astrophysical objects from each other. 64G246& attracts astrophysical objects to each other. 42G24& repels astrophysical objects from each other.

Gss.2.2.1   Stuff that constitutes the currently observable universe exited the period of inflation in a state such that people can consider that any curvature was adequately small that people can consider that zero- or negligible-curvature pertains at the time of exit or somewhat thereafter.

Part 3   Mathematics

Gss.3.2.1   For an edge case with -2ν an even positive integer, 1 type-1 solution exists.

Gss.3.2.2   For an edge case with -2ν an odd positive integer, 3 orthogonal type-1 solutions exist.

Gss.3.3.1   The number of generators, 48, for SU(7) correlates with some limits on solutions that correlate with elementary particles.

Gss.3.4.1   Some non-traditional solutions having $D_P=3$ correlate with the elementary particles, with some properties of elementary particles, and with fields related to elementary particles.

Gss.3.4.2   For solutions that correlate with non-zero-mass elementary particles, $\nu=-1$ correlates with elementary bosons and their fields, $\nu=-3/2$ correlates with elementary fermion particles, and $\nu=-1/2$ correlates with fermion fields.

Gss.3.4.3   For elementary particles, $\Omega=+S(S+1)>0$ correlates with QE-like phenomena, $\Omega=0$ correlates with the Higgs boson, and $\Omega=-S(S+1)<0$ correlates with QP-like phenomena.

Gss.3.4.4   For a solution to correlate with an elementary fermion, the relevant $\nu=-3/2$ (that is, particle) solution must correlate with a $\nu=-1/2$ (that is, field) solution.

Gss.3.4.5   For a solution to correlate with non-zero-mass elementary particles, D for the particles must be a positive integer.

Gss.3.4.6   For a solution to correlate with non-zero-mass elementary particles, D for the corresponding field must be a positive integer.

Gss.3.4.7   Regarding solutions that might qualify as correlating with elementary particles, we exclude solutions for which the number of instances is not an even divisor of 48.

Gss.3.4.8   For L-family representations, the number of -1 values correlating with QE-like oscillators plays a role, similar to the roles played by numbers of @ values for bosons, regarding determining numbers of instances.

Gss.3.4.9   People can overlook a possibility that the number of instances for 1L Dirac fermions is 48.

Gss.3.4.10   For any values of j and %, to the extent multiple values of k could pertain, any jkG%& solution that correlates with an elementary particle correlates with the minimal positive value of k. Here, k=2S.

Gss.3.4.11   The set of elementary particles includes 8 instances of gravitons. Each instance of gravitons correlates with a set of elementary fermions. Each such set of elementary fermions is identical to each other set.

Gss.3.4.12   O-family bosons with spin-S are bound into groups by Y-family members with spin-S.

Gss.3.4.13   The minimum number of O-family bosons in a bound group containing a (2S)O boson is 2S+1.

Gss.3.4.14   For S" a positive integer and S=S"$-(1/2)$, either all of the (2S")O, (2S")Y, and (2S)Q solutions correlate with elementary particles or none of those solutions correlates with an elementary particle.

Gss.3.4.15   The number of instances of (2S)Q equals the number of instances of (2S")Y, for S=S"$-(1/2)$.

Gss.3.4.16   The number of instances of (2S)Q equals the number of instances of (2S")O, for S=S"$-(1/2)$.

Gss.3.4.17   No 2kG%& force mediates an interaction between a fermion in MP(n) and a fermion in MP(n').

Gss.3.4.18   No 4kG%& force mediates an interaction between a fermion in MP(n) and a fermion in MP(n').

Gss.3.4.19   6kG%& forces can mediate interactions between fermions in MP(n) and fermions in MP(n'), for appropriate choices of n and n'.

Gss.3.4.20   8kG%& forces can mediate interactions between fermions in MP(n) and fermions in MP(n'), for appropriate choices of n and n'.

Gss.3.4.21   Items (3.4.147), (3.4.148), and (3.4.149) define, in effect, the term appropriate in items (3.4.144) and (3.4.145).

Gss.3.4.22   The number of instances of 0H0 is 1.

Gss.3.5.1 $\beta' = \beta$.

Gss.3.5.2   In the expression $(4/3)(\beta^6)^2 = \{(q_e)^2/(4\pi\varepsilon_0)\} / \{G_N(m_e)^2\}$, the leftmost exponent 2 represents the number of vertices in a Feynman diagram, β^6 represents the ratio of strengths per channel for electromagnetism and gravity (for interactions between 2 electrons), 4 represents the number of channels for a photon, and 3 represents the number of channels for a graviton.

Gss.3.5.3   We attach significance to R_j for which a particle property has an exponent k=0.

Gss.3.5.4   Regarding R_0, people can consider q_e to be a particle property for which $|q_e|^0$ pertains.

Gss.3.5.5   The U(1) in the W-family symmetry SU(3)×SU(2)×U(1) correlates with PR22G2 being (across elementary particles) positive for some elementary particles and being negative for some elementary particles.

Gss.3.5.6   PR66G6 can be positive or zero (but not negative), depending on the elementary particle or object being characterized.

Gss.3.5.7   PR88G8 can be positive or possibly zero (but not negative), depending on the elementary particle or object being characterized.

Gss.3.7.1   A merger of 2 sets (1 QE-like and 1 QP-like), each of 2 operators, correlates with a standard representation for $E^2-c^2P^2$ that people use based on the Dirac equation and the Klein-Gordon equation. Operators act on 4-component spinors. People can represent aspects of the operators via gamma matrices.

Gss.3.7.2   For Majorana neutrinos, wave-equations involving $E^2-c^2P^2$ involve 2-component spinors.

Gss.3.7.3   Z bosons, the 0H0 boson, and (2S)O0 bosons do not directly participate in interactions that change properties (such as charge) of fermions.

Gss.3.7.4   3Q fermions do not interact directly with 22G2& (photons).

Part 4   Invariant and relative properties

Gss.4.3.1   The confluence of at least 2 G-family series {out of 3 e-family series - the series of N(P0)=−1 bosons (44G4& and 22G2&), the series of #E=0 bosons (84G2468&, 64G246&, 42G24&, and 22G2&), and the j2G2& series (82G2&, 62G2&, 42G2&, and 22G2&)} at 22G2& correlates with the role of ℏ in the uncertainty principle.

Gss.4.4.1   For the L-family, combinations of the 4 solutions correspond to 2 of the 3 possible members of an L=1 set (the M=0 member does not pertain) and to the 1 member (M=0) of each of 2 L=0 sets.

Gss.4.4.2   For the Q-family, for S=1/2, combinations of the 4 solutions correspond to 4 of the 5 members of an L=2 set (the M=0 member does not pertain).

Gss.4.4.3   For charged leptons (either M'=−3 or M'=+3), people can benefit by correlating the range −1≤M''≤3 with an L=2 system.

Gss.4.4.4   For the L=2 system (with fixed M' and varying M'') that includes charged leptons and for some number ζ'', $m(M'',-3) \propto e^{M''\zeta''}(1+d(M''))$, in which −1≤M''≤3, d(0)=d(2), and d(−1)=d(1)=d(3)=0.

Gss.4.4.5   Regarding masses and charges for lepton siblings BMS3(2j), for 2j = 0, 2, or 4, m(3×2j,−3) = $(\beta')^{2j}×m(0,-3)$ and |Q'(3×2j,−3)| = |Q'(0,−3)|.

Gss.4.5.1   Quantum mechanically, the 2 sets of coordinates Table 1.1.6 presents suffice as a basis for modeling G-family #E=0 and #E=2 elementary-particle kinematics across all MP(n).

Part 5   Dark energy, dark matter, and sibling-state fermions

Gss.5.2.1   For each j'≥2, BMS'(j') states are unstable.

Part 6   New models for phenomena for which people say traditional theories correlate

Gss.6.3.1   Our approach exhibits conservation of angular momentum.

Gss.6.3.2   Our approach exhibits conservation of energy and conservation of momentum.

Gss.6.3.3   Within the physics of 1 instance of MP(n), people can consider that the W-family ground state and the 22G2& ground state have similarities.

Gss.6.5.1   All G-family physics can be discussed based on the flat space-time coordinates Table 1.1.6 shows.

Part 7   Models that may correlate with phenomena

Gss.7.1.1   Inflation correlates with conditions such that at least 1 of the following featured free particles - 1Q and 3Q.

Gss.7.8.1   ξ' = 0 correlates with C-mass=0 for G-family bosons.

Part 8   Other topics

Gss.8.2.1   People might think of the excitation of a G-family boson as including a step in which oscillator P0 is excited −N(P0) times and the QE-like oscillator pair for 1 channel is also excited −N(P0) times. Another step involves the P0 excitements being distriuted to Pjk oscillators (with each j≥2 and each k being either R or L) and the QE-like channel excitements being transferred to oscillator E0.

Gss.8.2.2   The expression $(3/2)(\beta^{(j×6)})^2$ could represent a ratio of the strengths of 22G2& to 66G6& for interactions between 2 electrons. Here, j might be 2.

Gss.8.2.3   The expression $(2/1)(\beta^{(k×6)})^2$ could represent a ratio of the strengths of 22G2& to 88G8& for interactions between 2 electrons. Here, k might be 3 or 4.

Gss.8.2.4   PR66G6 correlates with magnetic moment.

Gss.8.2.5   PR88G8 correlates with 2S.

Gss.8.5.1   People would say that statements above in this section correlate with our work's correlating with the SU(3)×SU(2)×U(1) symmetry people associate with the Standard Model.

Gss.8.5.2   People can extend work in this paper to include (or merge with) the Standard Model.

Part 9   Conclusions and extensions

Part 10   Compendia

Section 10.3   Possible opportunities for research

Here, we collect concepts correlating with possible opportunities for research.

Part 0   Introduction and summary

Tbd.0.5.1   For each statement following a ▪, to what extent does the statement point to possible opportunities for research?

Tbd.0.5.2   For each statement labelled Gss, to what extent does the statement point to possible opportunities for research?

Part 1   Elementary particles (not including dark matter or dark energy)

Part 2   Models that correlate with phenomena

Tbd.2.6.1   Verify or rule out (to some confidence level) the value for m_{tauon} we predict.

Tbd.2.6.2   Verify or rule out (to some confidence level) that $\beta=\beta'$.

Part 3   Mathematics

Tbd.3.4.1   Describe 3Q properties that correlate with each of oscillator pairs E4R and E4L, E2R and E2L, and P4L and P4R.

Tbd.3.4.2   Determine the group GRX for which the QE-like symmetry GRX×U(1) pertains for 4Y particles.

Tbd.3.6.1   To what extent does nature exhibit a set of particles (for each peer) correlating with the collection 4O, 3Q, and 4Y?

Tbd.3.6.2   Assuming 4O, 3Q, and 4Y particles exist, to what extent do 4O bosons interact directly with 1Q fermions?

Tbd.3.6.3   Assuming 4O, 3Q, and 4Y particles exist, to what extent do 2O and 2W bosons interact directly with 3Q fermions?

Tbd.3.6.4   Assuming that (for 1 peer) more than 1 sibling exists and assuming that multiple instances of 4O, 3Q, and 4Y pertain (for that 1 peer), to what extent do siblings and instances of 4O, 3Q, and 4Y correlate?

Tbd.3.6.5   To what extent can observations or experiments detect, infer, or rule out (to some confidence level) S>2 G-family bosons (such as might correlate with solutions 66G6&, 86G6&, and 88G8&)?

Part 4   Invariant and relative properties

Tbd.4.3.1   To what extent is there significance that, in formulas pertaining to a series of lengths [Table 3.5.1], G_N^{-1} correlates with c^0 and G_N^{+1} correlates with $\hbar^0$?

Tbd.4.4.1   Determine the extent to which the 2 S'=1/2 rows of Table 3.7.3 correlate with item (4.4.10).

Tbd.4.4.2   To what extent might people find significance in the relationship $(\tan^{-1}(1+d'') - 2^{-1/2}) / 2^{-1/2} \sim 4.6\times10^{-4}$?

Tbd.4.4.3   How might observations detect effects of BMS'(j'>0) zero-charge leptons?

Tbd.4.4.4   Try to create BMS'(j'>0) zero-charge leptons.

Tbd.4.4.5   Effect, detect, infer, or rule out (to some confidence level) the populating of the BMS'(1) M"=−2 zero-charge state in situations for which temperatures correlate with at least (or for which energies exceed) 2.5×10^3 eV.

Tbd.4.4.6   Effect, detect, infer, or rule out (to some confidence level) the populating of the BMS'(1) M"=+1 zero-charge state in situations for which temperatures correlate with at least (or for which energies exceed) 14 GeV.

Tbd.4.4.7   Effect, detect, infer, or rule out (to some confidence level) the populating of BMS'(2) M"=+2 zero-charge states in situations for which temperatures correlate with at least (or for which energies exceed) 1.06×10^2 MeV.

Tbd.4.4.8 Effect, detect, infer, or rule out (to some confidence level) the populating of the BMS'(3) M"=+6 zero-charge state in situations for which energies exceed (or for which temperatures correlate with at least) 4.5×10^3 GeV.

Tbd.4.4.9 To what extent to do solutions based on mathematics herein correlate with masses (or mass-math eigenvalues) for zero-charge leptons?

Tbd.4.4.10 Mathematically, to what extent does the lack of M"=1 charged leptons correlate with the existence of an M"=1 Majorana zero-charge lepton solution?

Tbd.4.4.11 Mathematically, to what extent does the muon mass not matching $\beta^{2/3}m_e$ [Item (4.4.70)] correlate with the existence of at least 1 of the M"=2 zero-charge lepton solutions?

Tbd.4.4.12 Effect, detect, infer, or rule out (to some confidence level) the populating of BMS6(j>0) non-zero-charge lepton states.

Tbd.4.4.13 Try to detect or create 3Q fermions.

Tbd.4.4.14 Try to detect or create compound particles that include 3Q fermions.

Tbd.4.4.15 To the extent S=3/2 elementary fermions exist, determine masses for S=3/2 elementary fermions.

Tbd.4.4.16 To the extent S=3/2 elementary fermions exist, determine masses for nucleon-like clusters of S=3/2 elementary fermions.

Tbd.4.4.17 To the extent S=3/2 elementary fermions exist, determine masses for nuclei-like clusters of nucleon-like clusters of S=3/2 elementary fermions.

Part 5   Dark energy, dark matter, and sibling-state fermions

Part 6   New models for phenomena for which people say traditional theories correlate

Tbd.6.4.1 To what extent might work in photonics benefit from people distinguishing phenomena related to relative (photon) quantities from phenomena related to invariant (photon) quantities?

Part 7   Models that may correlate with phenomena

Tbd.7.3.1 To what extent do photons with orbits that lie partly (but not wholly) outside a black hole's event horizon provide bases for black-hole thermal radiation and/or black-hole dissipation?

Tbd.7.4.1 A model that appropriately considers Q-mass and C-mass dovetails numerically with observed perihelion shifts.

Tbd.7.4.2 Determine the extent to which a model that considers Q-mass and C-mass (and possibly forces beyond Standard Model forces and gravity) dovetails with observed phenomena (that people associate with general-relativistic effects) that we do not address in this paper.

Tbd.7.4.3 Determine the extent to which a model that considers Q-mass and C-mass (and possibly forces beyond Standard Model forces and gravity) predicts observable yet-to-be-observed phenomena.

Tbd.7.5.1 For each of various galaxies, to what extent does cohesion depend on each of gravity as traditionally interpreted, the gravitational analog to electromagnetic magnetic fields, the presence of dark matter, and other phenomena?

Tbd.7.9.1 To what extent might people use O-family physics to refine, parallel, or supplant models people use regarding the nuclear force and/or nuclear physics?

Tbd.7.9.2 To what extent might people use mathematical models (that this paper includes) to refine, parallel, or supplant models for nuclear physics?

Tbd.7.9.3 To what extent might people use mathematical models (that this paper includes) and/or O-family physics to refine, parallel, or supplant models for neutron stars?

Part 8   Other topics

Tbd.8.2.1 Determine the f'(4), f'(6), j [Item (8.2.20)], k [Item (8.2.21)], and f'(8) that pertain to the magnitudes of G-family #E=0 forces.

Tbd.8.2.2 Determine strengths and directions (attraction or repulsion) for G-family forces other than 22G2& and 44G4&.

Tbd.8.2.3 Develop a suitable perturbation theory (possibly based on something like Feynman diagrams) for G-family interactions.

Tbd.8.2.4 Assuming that PR66G6 correlates with magnetic moment, determine the extent to which people can use such a perturbation theory [Item (8.2.41)] to estimate magnetic-moment anomalies. [Items (10.5.14) and (10.5.15)].

Tbd.8.2.5 To what extent does $\beta^{1/3} \approx e^e$ correlate with a symmetry (or other concept) that people would find meaningful?

Tbd.8.2.6   To what extent does $|\beta^{1/3} - e^e| \neq 0$ correlate with a mixing angle that correlates with the 66G6&, 86G6&, and 88G8& solutions and 2 instances of SU(3) symmetry?

Tbd.8.6.1   Provide theory correlating with tauon and muon decay rates.

Tbd.8.6.2   Pinpoint important interaction vertices pertaining to neutrino oscillations.

Tbd.8.6.3   Try to address the topic of arrow of time.

Tbd.8.6.4   To what extent does the existence of the 0H0 particle correlate with the existence of elementary particles with non-zero mass?

Tbd.8.6.5   To the extent work in this paper provides a complete list of elementary particles, to what extent might people say that nature does not exhibit magnetic monopoles?

Tbd.8.6.6   Look for, based on work above, possibly useful physics for bases to develop new observational approaches to infer or experiments to detect gravitons or gravitational waves.

Tbd.8.6.7   To what extent should people correlate the usefulness of space-time coordinates with usefulness for concepts of space time? To what extent can people describe nature (or, at least, quantum phenomena) without assuming space time has properties? To what extent can people describe nature (or, at least, quantum phenomena) without assuming space time exists?

Part 9   Conclusions and extensions

Tbd.9.0.1   Develop simpler bases for and/or descriptions of work this paper describes.

Tbd.9.1.1   Develop practical applications of harmonic-oscillator math beyond traditional applications and applications we discuss.

Tbd.9.1.2   Explore a broader range of solutions for a broader interpretation of the term multi-dimensional harmonic-oscillator math.

Tbd.9.2.1   To what extent can people use quantum-oscillator-based math to describe phenomena for which people use classical physics, electrodynamics, traditional quantum mechanics, the Standard Model, quantum chromodynamics, and/or general relativity?

Tbd.9.2.2   To what extent can people use results based on quantum-oscillator-based math to substitute quantum-based approximations for classical-physics-based approximations in simulations (for example, simulations of interactions between molecules; or, other simulations for which people might say that efforts span classical-physics/quantum-physics boundaries)?

Tbd.9.2.3   To what extent can people use quantum-oscillator-based math reproduce Standard Model math?

Tbd.9.2.4   To what extent, if any, does a quantum-oscillator-based approach require use of concepts people might term properties of space time?

Tbd.9.2.5   For phenomena (such as perihelia shifts) we describe qualitatively, develop quantitative treatments that match or predict observations.

Tbd.9.3.1   Detect or rule out (to some confidence level) various elementary particles to which Table 9.3.1 alludes.

Tbd.9.3.2   Estimate and/or measure quantities related to phenomena involving spin-3/2 fermions.

Tbd.9.3.3   Measure more accurately the tauon mass (m_{tauon}) and the gravitational constant (G_N); thereby, help clarify validity of theory we propose.

Tbd.9.3.4   Determine the extent to which distinguishing, for gravitons, between Q-mass-related phenomena and C-mass-related phenomena can lead to new tests for the existence and properties of gravitons.

Tbd.9.4.1   Verify, detect or rule out (to some confidence level) aspects to which Table 9.4.1 alludes.

Tbd.9.5.1   Verify, detect or rule out (to some confidence level) aspects to which Table 9.5.1 alludes.

Part 10   Compendia

Section 10.4   Names of tables

Here, we collect a list of table names.

Part 0   Introduction and summary
Table 0.5.1   Sections that collect lists
Part 1   Elementary particles (not including dark matter or dark energy)
Table 1.0.1   Parameters that correlate with properties of elementary particles

Thomas.J.Buckholtz@gmail.com   Copyright (c) 2014 Thomas J. Buckholtz   http://ThomasJBuckholtz.wordpress.com

Section 10.5   Some physics numbers

Here, we collect some numbers we use above.

The next table shows results of experiments. For items (10.5.14) and (10.5.15), $g_S=2$. [Ref.10.5.1, Ref.10.5.2, and Ref.10.5.3]

Table 10.5.1 Some physics numbers

Symbol	Units	Number	Description	
				(10.5.1)
q_e	C	$-1.602176565(35)\times10^{-19}$	charge of an electron	(10.5.2)
ε_0	F m^{-1}	$8.854187817\times10^{-12}$	permittivity of free space	(10.5.3)
G_N	m^3 kg^{-1} s^{-2}	$6.67545(18)\times10^{-11}$	gravitational constant	(10.5.4)
m_e	kg	$9.10938291(40)\times10^{-31}$	mass of an electron	(10.5.5)
m_e	MeV/c^2	$0.510998928(11)$	mass of an electron	(10.5.6)
$\hbar$	J s	$1.054571726(47)\times10^{-34}$	Planck constant, reduced	(10.5.7)
$\hbar$	MeV s	$6.58211928(15)\times10^{-22}$	Planck constant, reduced	(10.5.8)

Symbol	Units	Number	Description	(10.5.1)
c	m s^{-1}	2.99792458×10^8	speed of light in a vacuum	(10.5.9)
α	(no units)	$7.2973525698(24) \times 10^{-3}$	fine-structure constant	(10.5.10)
α^{-1}	(no units)	$137.035999074(44)$		(10.5.11)
m_{tauon}	MeV/c^2	$1.77682(16) \times 10^3$	mass of a tauon	(10.5.12)
m_{muon}	MeV/c^2	$105.6583715 \pm 0.0000035$	mass of a muon	(10.5.13)
a	(no units)	$(1159.65218076 \pm 0.00000027) \times 10^{-6}$	electron magnetic moment anomaly $(g - g_S) / g_S$	(10.5.14)
a	(no units)	$(11659209 \pm 6) \times 10^{-10}$	muon magnetic moment anomaly $(g - g_S) / g_S$	(10.5.15)

~ ~ ~

Ref.10.5.1   T. Quinn et al, Improved Determination of G Using Two Methods, *Phys. Rev. Lett,* 111, 101102, 2013. (http://link.aps.org/doi/10.1103/PhysRevLett.111.101102)

Ref.10.5.2   J. Beringer et al. (Particle Data Group), *Phys. Rev. D86*, 010001 (2012). (http://pdg.lbl.gov/2012/reviews/rpp2012-rev-phys-constants.pdf)

Ref.10.5.3   J. Beringer et al. (Particle Data Group), *Phys. Rev. D86*, 010001 (2012). (http://pdg.lbl.gov/2012/tables/rpp2012-sum-leptons.pdf)

Section 10.6   References

Here, we collect a list of references from above.

Part 0   **Introduction and summary**
Ref.0.4.1   John R. Gribbin and Mary Gribbin, *Richard Feynman, A Life In Science*, Dutton, 1997, page 178.

Part 1   **Elementary particles (not including dark matter or dark energy)**

Part 2   **Models that correlate with phenomena**
Ref.2.1.1   N. Gnedin, Cosmological Calculator for the Flat Universe. (http://home.fnal.gov/~gnedin/cc/)
Ref.2.1.2   N. G. Busca, et. al., Baryon Oscillations in the Lyα forest of BOSS quasars, arXiv:1211.2616 [astro-ph.CO].
Ref.2.2.1   NASA, http://map.gsfc.nasa.gov/universe/uni_shape.html

Part 3   **Mathematics**
Ref.3.2.1   Wolfram Alpha, computational knowledge engine, Wolfram Alpha LLC, http://mathworld.wolfram.com/DeltaFunction.html.
Ref.3.5.1   G. T. Adylov, et. al., A measurement of the electromagnetic size of the pion from direct elastic pion scattering data at 50 GeV/c, *Nuclear Physics B*, Volume 128, Issue 3, 3 October 1977, pages 461-505. (http://dx.doi.org/10.1016/0550-3213(77)90056-6)
Ref.3.5.2   Particle Data Group, Electroweak (web page), *The Particle Adventure*, Lawrence Berkeley National Laboratory, http://www.particleadventure.org/electroweak.html.
Ref.3.7.1   Wolfram Alpha, computational knowledge engine, Wolfram Alpha LLC, http://mathworld.wolfram.com/Laplacian.html.

Part 4   **Invariant and relative properties**
Ref.4.2.1   J. Beringer et al. (Particle Data Group), *PR D86*, 010001 (2012) and 2013 partial update for the 2014 edition (URL: http://pdg.lbl.gov). (http://pdg.lbl.gov/2013/tables/rpp2013-sum-gauge-higgs-bosons.pdf)

Ref.4.2.2   CMS collaboration (2012). "Observation of a new boson at a mass of 125 GeV with the CMS experiment at the LHC". *Physics Letters B* 716 (1): 30–61. arXiv:1207.7235. Bibcode:2012PhLB..716...30C. doi:10.1016/j.physletb.2012.08.021.

Ref.4.2.3   ATLAS collaboration (2012). "Observation of a New Particle in the Search for the Standard Model Higgs Boson with the ATLAS Detector at the LHC". *Physics Letters B* 716 (1): 1–29. arXiv:1207.7214. Bibcode:2012PhLB..716....1A. doi:10.1016/j.physletb.2012.08.020.

Ref.4.4.1   J. Beringer et al. (Particle Data Group), *Phys. Rev. D86*, 010001 (2012).

Ref.4.4.2   S. Thomas, F. Abdalla, and O. Lahav, Upper Bound of 0.28 eV on the Neutrino Masses from the Largest Photometric Redshift Survey, *Phys. Rev. Lett. 105*, 031301, 2010. (http://arxiv.org/abs/0911.5291)

Ref.4.4.3   A. Melchiorri, Constraints on Neutrino Physics from Planck, European Space Agency, http://www.rssd.esa.int/SA/PLANCK/docs/eslab47/Session06_CMB_Cosmology_and_Fundamental_Physics/47ESLAB_April_04_17_30_Melchiorri.pdf.

Ref.4.4.4   K.A. Olive *et al.* (Particle Data Group), Chin. Phys. C38, 090001 (2014) (URL: http://pdg.lbl.gov) - "Number of Neutrino Types"

Ref.4.4.5   K.A. Olive *et al.* (Particle Data Group), Chin. Phys. C38, 090001 (2014) (URL: http://pdg.lbl.gov) - "Leptons"

Part 5   Dark energy, dark matter, and sibling-state fermions

Ref.5.1.1   Mark Peplow, Planck telescope peers into primordial universe, *Nature News*, Nature Publishing Group, March 21, 2013. (http://www.nature.com/news/planck-telescope-peers-into-primordial-universe-1.12658)

Part 6   New models for phenomena for which people say traditional theories correlate

Part 7   Models that may correlate with phenomena

Part 8   Other topics

Ref.8.1.1   J. Beringer et al. (Particle Data Group), *Phys. Rev. D86*, 010001 (2012). (http://pdg.lbl.gov/2012/reviews/rpp2012-rev-cosmic-microwave-background.pdf)

Ref.8.3.1   J. Beringer et al. (Particle Data Group), *PR D86*, 010001 (2012) and 2013 partial update for the 2014 edition (URL: http://pdg.lbl.gov).

Part 9   Conclusions and extensions

Part 10   Compendia

Ref.10.5.1   T. Quinn et al, Improved Determination of G Using Two Methods, *Phys. Rev. Lett,* 111, 101102, 2013. (http://link.aps.org/doi/10.1103/PhysRevLett.111.101102)

Ref.10.5.2   J. Beringer et al. (Particle Data Group), *Phys. Rev. D86*, 010001 (2012). (http://pdg.lbl.gov/2012/reviews/rpp2012-rev-phys-constants.pdf)

Ref.10.5.3   J. Beringer et al. (Particle Data Group), *Phys. Rev. D86*, 010001 (2012). (http://pdg.lbl.gov/2012/tables/rpp2012-sum-leptons.pdf)

Endnotes

(We reserve this area for updates, for notes, and/or for endnotes regarding updates we make to material we present before this area.)